Engineering Metamaterials

Engineering Metamaterials

Editor

Stanislav Maslovski

MDPI • Basel • Beijing • Wuhan • Barcelona • Belgrade • Manchester • Tokyo • Cluj • Tianjin

Editor
Stanislav Maslovski
Departamento de Eletrónica, Telecomunicações e Informática
Universidade de Aveiro
Campus Universitário de Santiago
3810-193 Aveiro, Portugal

Editorial Office
MDPI
St. Alban-Anlage 66
4052 Basel, Switzerland

This is a reprint of articles from the Special Issue published online in the open access journal *Electronics* (ISSN 2079-9292) (available at: https://www.mdpi.com/journal/electronics/special_issues/eng_metamater).

For citation purposes, cite each article independently as indicated on the article page online and as indicated below:

LastName, A.A.; LastName, B.B.; LastName, C.C. Article Title. *Journal Name* **Year**, *Article Number*, Page Range.

ISBN 978-3-03936-922-5 (Hbk)
ISBN 978-3-03936-923-2 (PDF)

Contents

About the Editor . . . . **vii**

Stanislav Maslovski
Engineering Metamaterials: Present and Future
Reprinted from: *Electronics* **2020**, *9*, 932, doi:10.3390/electronics9060932 . . . . **1**

Ayesha Kosar Fahad, Cunjun Ruan and Kanglong Chen
Dual-Wide-Band Dual Polarization Terahertz Linear to Circular Polarization Converters based
on Bi-Layered Transmissive Metasurfaces
Reprinted from: *Electronics* **2019**, *8*, 869, doi:10.3390/electronics8080869 . . . . **5**

Go Itami, Osamu Sakai and Yoshinori Harada
Two-Dimensional Imaging of Permittivity Distribution by an Activated Meta-Structure with a
Functional Scanning Defect
Reprinted from: *Electronics* **2019**, *8*, 239, doi:10.3390/electronics8020239 . . . . **19**

Francesca Venneri, Sandra Costanzo and Antonio Borgia
A Dual-Band Compact Metamaterial Absorber with Fractal Geometry
Reprinted from: *Electronics* **2019**, *8*, 879, doi:10.3390/electronics8080879 . . . . **35**

Myunghoi Kim
Analytical Modeling of Metamaterial Differential Transmission Line Using Corrugated Ground
Planes in High-Speed Printed Circuit Boards
Reprinted from: *Electronics* **2019**, *8*, 299, doi:10.3390/electronics8030299 . . . . **43**

**Shan Yin, Xintong Shi, Wei Huang, Wentao Zhang, Fangrong Hu, Zujun Qin and
Xianming Xiong**
Two-Bit Terahertz Encoder Realized by Graphene-Based Metamaterials
Reprinted from: *Electronics* **2019**, *8*, 1528, doi:10.3390/electronics8121528 . . . . **57**

**Vasa Radonić, Slobodan Birgermajer, Ivana Podunavac, Mila Djisalov, Ivana Gadjanski and
Goran Kitić**
Microfluidic Sensor Based on Composite Left-Right Handed Transmission Line
Reprinted from: *Electronics* **2019**, *8*, 1475, doi:10.3390/electronics8121475 . . . . **67**

Kimberley W. Eccleston and Ian G. Platt
Identifying Near-Perfect Tunneling in Discrete Metamaterial Loaded Waveguides
Reprinted from: *Electronics* **2019**, *8*, 84, doi:10.3390/electronics8010084 . . . . **81**

About the Editor

Stanislav Maslovski (Dr., Assoc. Research Prof.) was born in Leningrad, U.S.S.R. (presently St. Petersburg, Russia), in 1975. He received his B.Sc., M.Sc., and Cand. Math-Phys. Sc. (Ph.D.) degrees from the Radiophysics faculty of St. Petersburg State Technical Univ. (SPb STU) in 1997, 1999 and 2004, respectively. During the years 2002–2005, he was with the Radio Laboratory of Helsinki Univ. of Technology. In the period 2006–2008, S.I. Maslovski worked as a Docent (Assoc. Prof.) at the Radiophysics faculty of SPb STU. Since 2009, he has been with Instituto de Telecomunicações (IT), Portugal. During the years 2010–2013, S.I. Maslovski was also a visiting Senior Research Fellow at the Laboratory of Metamaterials of the National Research University of Information Technologies, Mechanics and Optics, St. Petersburg, Russia. Presently, S.I. Maslovski is Associate Research Professor at Dept. of Electronics, Telecommunications and Informatics at Aveiro Univ., Portugal, and Senior Researcher at IT-Aveiro. He is an Associate Editor in IET Quantum Communication and a Topic Editor in Electronics (Switzerland). His main research interests are in the electromagnetics of metamaterials and metasurfaces, smart and active antennas and beamforming systems, and in the quantum-electromagnetic effects in metamaterials, such as the Casimir–Lifshitz forces and the super-Planckian radiative heat transfer effects.

Editorial

Engineering Metamaterials: Present and Future

Stanislav Maslovski

Instituto de Telecomunicações and Departamento de Eletrónica, Telecomunicações e Informática, Universidade de Aveiro, Campus Universitário de Santiago, 3810-193 Aveiro, Portugal; stas@av.it.pt

Received: 26 May 2020; Accepted: 29 May 2020; Published: 4 June 2020

1. Introduction

A couple of decades have passed since the advent of electromagnetic metamaterials. Although the research on artificial microwave materials dates back to the middle of the 20th century, the most prominent development in the electromagnetics of artificial media has happened in the new millennium. In the last two decades, the electromagnetics of one-, two-, and three-dimensional metamaterials acquired robust characterization, measurement, and design tools (e.g., [1–8]). Novel fabrication techniques have been developed. Many exotic effects involving metamaterials and metasurfaces, which initially belonged in a scientist's lab, are now well understood by practicing engineers. Therefore, when accepting the Guest Editor role for this Special Issue, I decided that it was the right moment to bring up and refresh the metamaterial concepts, which had to become a designer's tools of choice in the present-day electronics, microwaves, and photonics.

Time had shown that I made a step in the right direction. The papers published in this Special Issue had covered several important topics advancing the state of the art in telecommunications, subwavelength imaging, and biomedical sensing. Although some of these works build up on the metamaterial- and metasurface-related studies done in the last decade or even earlier, the originality in the selected papers resides in the approaches, models, and experimental techniques that demonstrate a great value for practical applications. That explains one of the reasons why this Special Issue has been entitled "Engineering Metamaterials". However, there is another reason: practical applications require from us, scientists and engineers, to search for metamaterial realizations that are compatible with the present-day technologies and respond to requests from foreseable future. That is why the engineers must be able to not just use, but also *design and engineer* metamaterials for a specific need. I believe, the reader will find this Special Issue useful for that purpose.

2. The Present Issue

The focus of this Special Issue has been on the theory and applications of electromagnetic metamaterials, metasurfaces, and metamaterial transmission lines as the building blocks of present-day and future electronic, photonic, and microwave devices. Below, I outline the main results obtained in the papers published in this Special Issue.

In Reference [9], the authors have studied the near-perfect tunneling in discrete metamaterial loaded waveguides. Electromagnetic wave tunneling in alternating ε-negative/μ-negative anisotropic metamaterial layers is studied with an ABCD-matrix-based method. A tunnel identification method is developed and demonstrated to reveal tunneling behavior experimentally. Two-dimensional imaging of permittivity distribution by an activated meta-structure with a functional scanning defect is studied in Reference [10]. A perforated metal plate with subwavelength holes and a needle-like conductor are employed to perform two-dimensional scanning. Imaging experiments have been conducted with the aim of detecting both conductive and dielectric samples for future biomedical applications concerning non-invasive and low-risk diagnosis. In Reference [11], analytical modeling of metamaterial differential transmission line using corrugated ground planes in high-speed printed circuit boards is performed.

The developed model enables efficient and accurate prediction of the common-mode noise suppression and differential signal transmission characteristics. The authors of Reference [12] investigate wide-band dual polarization terahertz linear to circular polarization converters based on bi-layered transmissive metasurfaces. The bi-layered metasurfaces are formed by diagonally intersecting square metallic patches and rings. An equivalent circuit model is developed and validated with full-wave simulations. Reference [13] deals with a dual-band compact metamaterial absorber with fractal geometry. The used fractal structure allows for creating a dual-band metamaterial absorber with reduced unit cell size and small substrate thickness. Based on this principle, an absorber panel has been experimentally realized and validated. In Reference [14], a microfluidic sensor based on a composite left-right handed transmission line has been proposed and investigated. A change in the properties of the fluid that fills the microfluidic reservoir causes a change in the effective substrate permittivity, which subsequently changes the phase velocity in the transmission line. The studied sensor is characterized by relatively high sensitivity and good linearity and can be used for the biomass estimation inside microfluidic bioreactors. Two-bit terahertz encoder realized by graphene-based metamaterials is proposed and studied in Reference [15]. The encoder involves graphene-based metamaterials, in which the graphene structures are controlled by electric voltage applied to external electrodes. The authors foresee that their encoder can promote the development of multifunctional and integrated devices for future THz-band communications.

3. Future

Rapid growth in telecommunications and electronics requires development of new and original metamaterial-based and (or) metamaterial-inspired devices. Recently, digital metamaterials and programmable metasurfaces have attracted a lot of attention. Programmable metasurfaces are versatile tools for controlling electromagnetic wave propagation and performing almost instantaneous operations on the wavefronts of passing electromagnetic waves. It is without doubt that the metamaterial concepts (including the ones presented in this Special Issue) will be further developed for such applications.

Telecommunications in the mm-wave and THz bands will bring new challenges that will be addressed, in particular, by making use of materials with exotic electronic properties such as the graphene. In the last few years, topological metamaterials and effects have added an entirely new dimension to the whole picture. Emerging artificial intelligence techniques already assist researchers in design of novel metamaterials and devices based on them. Overall, I believe that there is a bright future for these new technologies and their applications in *engineering metamaterials*.

Acknowledgments: Here, first of all, I would like to thank all the Special Issue authors for their valuable contributions and acknowledge the work of anonymous reviewers and the editorial board of *Electronics*. Without your effort this Special Issue could not become a reality! The work related to Guest Editing of this Special Issue and publication of this Editorial was funded by FCT/MCTES, Portugal, through national funds and when applicable co-funded by EU funds under the project UIDB/50008/2020-UIDP/50008/2020.

Conflicts of Interest: The author declares no conflict of interest.

References

1. Tretyakov, S.A.; Nefedov, I.S.; Simovski, C.R.; Maslovski, S.I. Modelling and Microwave Properties of Artificial Materials with Negative Parameters. In *Advances in Electromagnetics of Complex Media and Metamaterials, NATO Science Series (Series II: Mathematics, Physics and Chemistry)*; Zouhdi, S., Sihvola, A., Arsalane, M., Eds.; Springer: Dordrecht, The Netherlands, 2003; Chapter 89, pp. 99–122. [CrossRef]
2. Kärkkäinen, M.K.; Maslovski, S.I. Wave propagation, refraction, and focusing phenomena in Lorentzian double-negative materials: A theoretical and numerical study. *Microw. Opt. Technol. Lett.* **2003**, *37*, 4–7. [CrossRef]
3. Maslovski, S.; Tretyakov, S.; Alitalo, P. Near-field enhancement and imaging in double planar polariton-resonant structures. *J. Appl. Phys.* **2004**, *96*, 1293–1300. [CrossRef]

4. Alitalo, P.; Maslovski, S.; Tretyakov, S. Near-field enhancement and imaging in double cylindrical polariton-resonant structures: Enlarging superlens. *Phys. Lett. A* **2006**, *357*, 397–400. [CrossRef]
5. Maslovski, S.I.; Silveirinha, M.G. Nonlocal permittivity from a quasistatic model for a class of wire media. *Phys. Rev. B* **2009**, *80*, 245101. [CrossRef]
6. Costa, J.T.; Silveirinha, M.G.; Maslovski, S.I. Finite-difference frequency-domain method for the extraction of effective parameters of metamaterials. *Phys. Rev. B* **2009**, *80*, 235124. [CrossRef]
7. Maslovski, S.I.; Morgado, T.A.; Silveirinha, M.G.; Kaipa, C.S.R.; Yakovlev, A.B. Generalized additional boundary conditions for wire media. *New J. Phys.* **2010**, *12*, 113047. [CrossRef]
8. Luukkonen, O.; Maslovski, S.I.; Tretyakov, S.A. A Stepwise Nicolson–Ross–Weir-Based Material Parameter Extraction Method. *IEEE Antennas Wirel. Propag. Lett.* **2011**, *10*, 1295–1298. [CrossRef]
9. Eccleston, K.W.; Platt, I.G. Identifying Near-Perfect Tunneling in Discrete Metamaterial Loaded Waveguides. *Electronics* **2019**, *8*, 84. [CrossRef]
10. Itami, G.; Sakai, O.; Harada, Y. Two-Dimensional Imaging of Permittivity Distribution by an Activated Meta-Structure with a Functional Scanning Defect. *Electronics* **2019**, *8*, 239. [CrossRef]
11. Kim, M. Analytical Modeling of Metamaterial Differential Transmission Line Using Corrugated Ground Planes in High-Speed Printed Circuit Boards. *Electronics* **2019**, *8*, 299. [CrossRef]
12. Fahad, A.K.; Ruan, C.; Chen, K. Dual-Wide-Band Dual Polarization Terahertz Linear to Circular Polarization Converters based on Bi-Layered Transmissive Metasurfaces. *Electronics* **2019**, *8*, 869. [CrossRef]
13. Venneri, F.; Costanzo, S.; Borgia, A. A Dual-Band Compact Metamaterial Absorber with Fractal Geometry. *Electronics* **2019**, *8*, 879. [CrossRef]
14. Radonić, V.; Birgermajer, S.; Podunavac, I.; Djisalov, M.; Gadjanski, I.; Kitić, G. Microfluidic Sensor Based on Composite Left-Right Handed Transmission Line. *Electronics* **2019**, *8*, 1475. [CrossRef]
15. Yin, S.; Shi, X.; Huang, W.; Zhang, W.; Hu, F.; Qin, Z.; Xiong, X. Two-Bit Terahertz Encoder Realized by Graphene-Based Metamaterials. *Electronics* **2019**, *8*, 1528. [CrossRef]

Article

Dual-Wide-Band Dual Polarization Terahertz Linear to Circular Polarization Converters based on Bi-Layered Transmissive Metasurfaces

Ayesha Kosar Fahad [1], Cunjun Ruan [1,2,*] and Kanglong Chen [1]

[1] School of Electronic and Information Engineering, Beihang University, Beijing 100191, China
[2] Beijing Key Laboratory for Microwave Sensing and Security Applications, Beihang University, Beijing 100191, China
*   Correspondence: ruancunjun@buaa.edu.cn; Tel.: +86-1350-1205-336

Received: 7 June 2019; Accepted: 4 August 2019; Published: 6 August 2019

Abstract: Transmissive metasurface-based dual-wide-band dual circular polarized operation is needed to facilitate volume and size reduction along with polarization diversity for future THz wireless communication. In this paper, a novel dual-wide-band THz linear polarization to circular polarization (LP-to-CP) converter is proposed using transmissive metasurfaces. It converts incident X polarized waves into transmitted left-hand circular polarized (LHCP) and right-hand circular polarized (RHCP) waves at two frequency bands. The structure consists of bi-layered metasurfaces having an outer conductor square ring and three inner conductor squares diagonally intersecting each other. The proposed converter works equally well with incident Y polarizations. Operational bandwidths for the dual-band LP-to-CP are 1.16 THz to 1.634 THz (34% fractional bandwidth) and 3.935 THz to 5.29 THz (29% fractional bandwidth). The electromagnetic simulation was carried out in two industry-standard software packages, High Frequency Structure Simulator (HFSS) and Computer Simulation Technology (CST), using frequency and time domain solvers respectively. Close agreement between results depicts the validity and reliability of the proposed design. The idea is supported by equivalent circuits and physical mechanisms involved in the dual-wide-band dual polarization operation. The impact of different geometrical parameters of the unit cell on the performance of LP-to-CP operation is also investigated.

Keywords: metasurfaces; linear to circular polarization converter; dual-band polarization converters; transmission-based polarization conversion

---

1. Introduction

Terahertz (THz) electromagnetic waves ranging from 0.1 THz to 10 THz have been investigated in numerous applications including imaging, spectroscopy, environmental surveillance, remote sensing, high-resolution radars and high-speed communications [1–4]. With the advancement in THz sources and detectors, these investigations have gained great interest from researchers. Polarization manipulators, in their applications to rotate or convert polarization states of electromagnetic waves, have been comprehensively explored for stealth, cloaking and in diode-like applications [5–8]. Circular polarization is a preferred choice for THz wireless communication due to its lower sensitivity towards multipath fading and polarization mismatch with receiver mobility.

Dual-band circular polarized waves are required in any satellite communication for the uplink (U/L) and downlink (D/L) operational bands. The handedness for U/L and D/L operations needs to be opposite to offer adequate polarization diversity. A generic solution for this is to use a linearly polarized antenna in cascade with orthomode transducer (OMT) as a feeder to a transmit-array or reflector. This solution is quite complicated, bulky and expensive. Another solution is to use phased array dual-band patch antennas. Both of these solutions are not feasible at THz frequencies. There is another possibility to obtain dual-band dual polarizations, i.e., using a linearly polarized wave generated by a linearly polarized antenna in cascade with dual-band linear polarization to circular polarization (LP-to-CP) converters. This solution is advantageous in terms of size and complexity. In addition, dual-band operation for LP-to-CP can be used to merge multiple systems for cost and volume reduction.

In the past, birefringent structures have been used to convert one state or type of polarization into another, including waveplates [9–11], liquid crystals [12–14], and wood and paper [15,16]. However, these solutions are bulky and complicated to integrate with existing THz systems. Metamaterials, chiral metamaterials and quasi-periodic planar arrays of sub-wavelength elements have attracted many researchers because of their distinguished properties, such as asymmetric transmission and polarization conversion with tenability and flexibility [17–19]. Metasurfaces, as the 2D equivalent of metamaterials, have been explored for the possibility of polarization manipulation, including linear to circular polarization conversion [20–22]. Particularly, they have been explored for single band linear to circular polarization conversion capability [20,23–27]. Wideband linear to circular polarization conversion operation has been explored using reflection and transmission modes [23–27]. Controlling electromagnetic fields, termed 'wave engineering', has been explored using metasurfaces [28–32]. Hadad et al. [28] proposed transverse temporal gradient-based metasurfaces for efficient transmission that is otherwise difficult using thin layered metasurfaces. Taravati et al. [29] proposed an advanced wave engineering technique based on unidirectional frequency generation and spatial decomposition in space–time-modulated slabs. Taravati et al. [30] realized an extraordinary beam splitter with one-way beam splitting-amplification. The proposed technique offers high isolation, transmission gain and zero beam tilting. Shi et al. [31] proposed a nonreciprocal metasurface that can achieve optical circulation and isolation. Wang et al. [32] proposed a technique for nonreciprocal wavefront engineering using time-modulated gradient metasurfaces. The essential building block of these surfaces is a subwavelength unit-cell whose reflection coefficient oscillates at low frequency. Such devices have been demonstrated theoretically [28–31] or experimentally [32] in an excellent manner for a wide range of applications, including cloaking, camouflage, amplifiers, isolators, duplexer antenna systems and mixers. However, there has been a lack of devices with incident normal polarization with transmitted circularly polarized waves using metasurfaces. Such devices are required in systems where transmitted waves and incident waves need to be aligned. These systems are predominantly used in THz wireless communication systems, including satellite communication. Such devices fabricated on flexible substrates can be integrated with linearly polarized wide-band THz antennas to have a dual wide-band outgoing transmitted wave with opposite handedness.

For dual-band LP-to-CP operation, there can be two possibilities: one is to use a wideband LP-to-CP converter so as to cover both required frequency bands. This is usually very difficult as it will practically increase the operational bandwidth for LP-to-CP operation. In the literature, a fractional bandwidth greater than 40% using transmissive metasurfaces has not been quoted so far. Moreover, no work has been reported to claim dual polarization in the same frequency band so as to cover applications requiring polarization diversity. The other possibility is to design a dual-band LP-to-CP converter that works over two different frequency bands with outgoing circular polarizations having opposite handedness in two bands.

Dual-band polarization manipulators including cross polarization converters [33,34] and LP-to-CP [35–39] converters have been proposed in the past. Liu [34] et al. and Xiaojun Huang et al. [35] reported metasurfaces based dual-band polarization converters for outgoing cross polarization. Recently reported dual band LP-to-CP polarization converters in microwave and THz bands are either complex or based on reflection type frequency selective surfaces [36–40]. Moreover, the operating bandwidths for the two bands are not wide. For example, Qingyun et al. [36] reported dual-band transmission type LP-to-CP converter based on frequency selective surfaces. Qingyun et al. [36] used four metallic layers to obtain 31.6% and 13.8% fractional bandwidths in C and Ku bands, respectively. The proposed structure's first and fourth metal layers consist of a split ring resonator bisected by a metallic strip; second and third metallic layers consist of a rectangular patch surrounded by a rectangular ring. Parinaz et al. [37] presented the design of dual-band LP-to-CP using transmissive metasurfaces whose unit cell is composed of three metallic layers. The first and third layers consist of a metallic patch enclosed in a split ring resonator, whereas the second layer consists of a circle-eliminated square patch with a central rectangular metallic strip. In other research, 3 dB fractional bandwidths of about 5% and 8% for dual-bands in the Ka band have been achieved [37]. Wang et al. [38] presented a dual-band LP-to-CP converter using Jerusalem cross and "I" dipole patterned frequency selective surfaces. Dual-band operation has been achieved for 29% and 12% fractional bandwidths. Youn et al. [39] reported a multilayered radial-shaped resonator-based dual-band LP-to-CP in the Ka band and achieved 14.4% and 4% fractional bandwidths. The only work reported of a THz dual-band LP-to-CP [40] discusses a reflection-based double-layered structure with the bottom layer used as gold reflectors. Such devices achieve a wide band of operation but tend to block and interfere with feeding elements such as linearly polarized arrays. The transmission-based dual-wide-band LP-to-CP converter based on a simple configuration is still a challenging problem.

In this paper, a novel dual-wide-band dual polarized LP-to-CP THz converter consisting of bi-layered transmissive metasurfaces has been presented. The proposed structure is novel, as it achieves dual-band widest transmission-based LP-to-CP operation with fractional bandwidths of 34% and 29% for the two bands in THz. Also, the structure consists of bi-layered metasurfaces. Dual-band operation is realized due to the excitation of two Eigenmodes generating phase delays. The position of frequency bands can be tuned by tuning the dimensions of square patches.

This paper is organized as follows: Section 2 of this paper describes the design of the proposed dual-wide-band LP-to-CP converter. Section 3 is about simulation and analysis of the dual-band LP-to-CP converter, and the principle of operation is described in Section 4. Physical mechanisms and equivalent circuit analyses are explained in Section 5. The impact of different structural parameters on the performance of the dual-band LP-to-CP converter is discussed in Section 6. The conclusion is presented in Section 7.

2. Design

Basic polarization manipulation properties of metasurfaces are achieved due to cross-coupling between electric and magnetic fields' resonances in the presence of an incident wave. In order to achieve dual-band LP-to-CP operation, the unit-cell structure of the metasurface needs to be designed deliberately to tailor the cross-coupling effects in two separate bands of interest. The same structure will behave differently under different frequencies. Each frequency band will correspond to a different Eigenmode. In the past, many diagonal symmetry/semi-symmetric anisotropic structures have been proposed for single band LP-to-CP conversion [41–45]. Dual band LP-to-CP operation has been achieved using center-connected [38] and semi-diagonal symmetric [37] structures. Symmetric structures with horizontal and vertical axis symmetry cannot be a choice for LP-to-CP operation because the electric response for such components at normal incidence and horizontal polarization will not generate a vertical component [42]. A good choice is to have a diagonal symmetric structure so that incident wave can be divided into two equal orthogonal components. The first design consideration for the unit cell is to have two differently sized square patches, with each square corresponding to a single band LP-to-CP operation. They are arranged to have diagonal symmetry. The second consideration is a square ring to have closed form electric field distribution between the top and bottom surfaces. In a nutshell, the proposed structure is a diagonal symmetric structure which consists of multiple resonating structures to have wider bandwidths in two operational bands. Figure 1 shows the design of the proposed converter. It consists of two identical sheets of conductor patches having substrate layer sandwiched between. Gold was used as a conductor and flexible polyimide having ε_r of 3.5 was used as a substrate. The shaded region in Figure 1b shows the conductor layer, which consists of two parts: the outer part is a square ring consisting of a conductor with width w and an inner part consisting of three squares diagonally intersecting each other. Top right and bottom left squares have dimensions $c \times c$, while the middle square has dimension $c1 \times c1$. The bottom layer is identical to the top layer. The design was optimized to have the best results and optimized parameters for the unit cell, as follows: $p = 50.67$ μm, $v = 54.65$ μm, $c = 13.4$ μm, $d = 3.5$ μm, $c1 = 14$ μm. The scheme for the operation of the dual-wide-band LP-to-CP converter is shown in Figure 2.

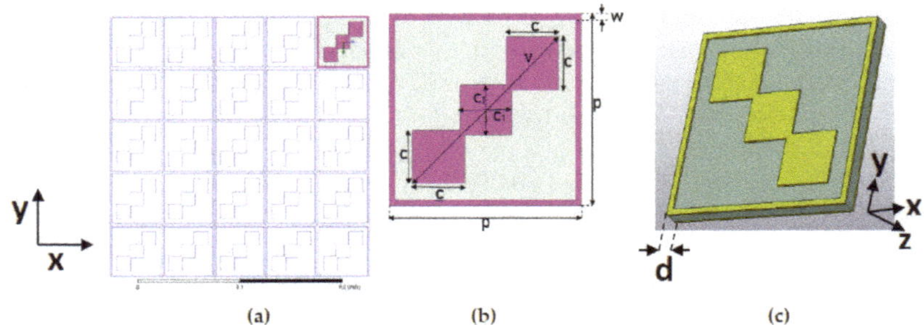

(a)   (b)   (c)

Figure 1. Schematic of the design: (**a**) Two-dimensional (2D) periodic array structure; (**b**) top view; (**c**) 3D perspective view.

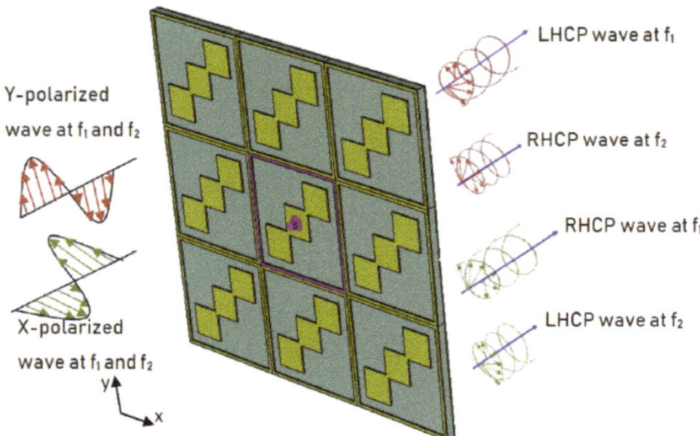

Figure 2. Scheme for dual-wide-band LP-to-CP converter with outgoing Left Handed Circular Polarization (LHCP) and Right Handed Circular Polarization (RHCP) waves.

3. Simulation and Analysis

Design and optimization for the proposed dual-band LP-to-CP converter was carried out using standard electromagnetic software, High Frequency Structure Simulator (HFSS). HFSS is based on the finite element mesh (FEM) solver. The design was simulated using master–slave boundary conditions, and Floquet ports at the input and output of the unit cell were applied to realize the periodic array structure. In order to validate the performance of the proposed structure, an optimized unit cell was re-simulated in Computer Simulation Technology (CST) software (2015 Version, Dassault Systèmes SE, Vélizy-Villacoublay, France). The finite difference time domain (FDTD) solver was selected for CST. Scanning time was set to 200 ns to get accurate results. Close agreement between FEM results and FDTD results validate the performance of the proposed structure. Total transmission in the X direction can be computed by $T_{all} = \left| t_{yx} \right|^2 + \left| t_{xx} \right|^2$ [46], and for transmission in the Y direction by $T_{all} = \left| t_{yy} \right|^2 + \left| t_{xy} \right|^2$ [46]. Figure 3a,b depict transmission characteristics for the incident X polarized and Y polarized wave travelling in the -Z direction in CST and HFSS. It is pertinent to mention here that transmission spectra with incident X and Y polarizations are not exactly equal due to the absence of symmetry in X and Y planes. Figure 4 shows the phase difference in degrees with incident X polarization. Further discussion on Figures 3 and 4 is carried out in Section 4.

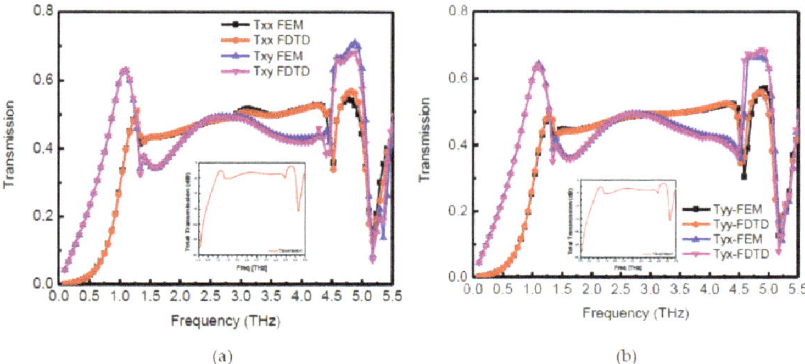

Figure 3. Transmission characteristics of proposed structure with incident: (**a**) X polarized wave, and (**b**) Y polarized wave.

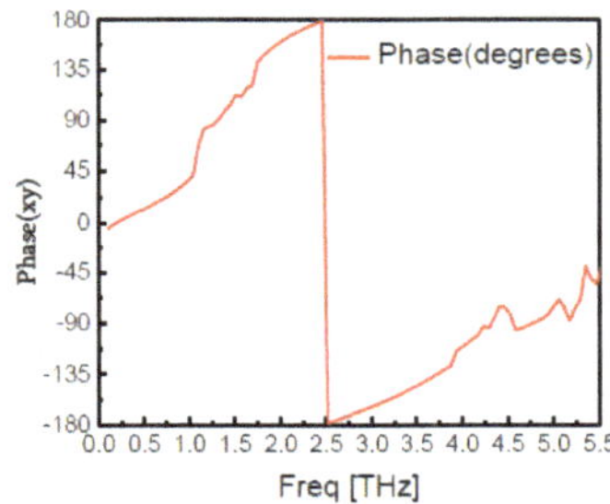

Figure 4. Phase difference between X polarized and Y polarized transmitted waves with incident X polarized wave.

4. Principle of Operation

In the topology diagram shown in Figure 2, it is assumed that both X polarized and Y polarized waves can be made incident on the metasurface. The proposed device has different responses to incident X and Y polarized waves. For incident X polarized waves, transmitted waves behave as LHCP at f_1 and RHCP at f_2, whereas for incident Y polarized waves, transmitted waves behave as RHCP at f_1 and LHCP at f_2. In order to understand the operation of a dual-band LP-to-CP converter, a plane horizontal (X polarized) wave travelling in the -Z direction is made incident on the surface of the unit cell. The incident wave can be expressed by Equation (1). Magnitudes of this incident wave can be expressed by Equation (2).

$$\mathbf{E_{xi}} = E_{xi}\mathbf{e_x} \tag{1}$$

$$\text{where, } E_{xi} = E_0 e^{jkz} \tag{2}$$

where $\mathbf{e_x}$ is the unit cell in X direction. The transmitted wave can be expressed as the sum of two components, i.e., X polarized and Y polarized, as shown in Equation (3):

$$\mathbf{E_t} = E_{xt}\mathbf{e_x} + E_{yt}\mathbf{e_y} = t_{xx}e^{j\varphi_{xx}}E_0 e^{jkz}\mathbf{e_x} + t_{xy}e^{j\varphi_{xy}}E_0 e^{jkz}\mathbf{e_y} \tag{3}$$

$$t_{xx} = \frac{E_{xt}}{E_{xi}} \tag{4}$$

$$t_{xy} = \frac{E_{yt}}{E_{xi}} \tag{5}$$

where t_{xx} and t_{xy} represent transmission coefficients for X to X and X to Y polarization conversion as shown in Equations (4) and (5), respectively. φ_{xx} and φ_{xy} are phase angles corresponding to t_{xx} and t_{xy}, respectively. Since the proposed structure has an anisotropic structure, the magnitudes and phasers for X polarized and Y polarized transmitted wave components may be different. However, if for a certain frequency range these transmission coefficients become comparable and their phase angles are 90° apart, i.e., $t_{xx} = t_{xy}$ and $\varphi_{xy} = 2n\pi \pm \pi/2$, with n being an integer, then the conditions for linear-to-circular polarization conversion will be met. In order to describe the transmission conversion performance of the proposed structure, the axial ratio for the transmitted wave is calculated as given in Equation (6) [38]:

$$AR = \left(\frac{\left|t_{xx}\right|2 + \left|t_{xy}\right|^2 + \sqrt{a}}{\left|t_{xx}\right|2 + \left|t_{xy}\right|^2 - \sqrt{a}} \right)^{1/2} \tag{6}$$

where a can be calculated from Equation (7) [38]:

$$a = \left|t_{xx}\right|^4 + \left|t_{xy}\right|^4 + 2\left|t_{xx}\right|^2\left|t_{xy}\right|^2\cos\left(2\varphi_{xy}\right) \tag{7}$$

For an ideal LP-to-CP operation, AR should be 1 (0 dB). However, for most systems, a 3-dB value of the axial ratio is acceptable.

Figures 3a and 4 show that in frequency bands from 1.16 THz to 1.634 THz and 3.935 THz to 5.29 THz, the transmission coefficient magnitudes are comparable, and the phase difference between them is −90° or +270° with ±15° variation. Thus, the conditions for linear to circular polarization conversion is fully met at some frequencies, while for a range of frequencies, it partially fulfils the requirements (in this case the transmitted wave will be slightly elliptically polarized). Nonetheless, the performance criterion for linear to circular transmission type conversion (axial ratio within 3 dB) is maintained. Furthermore, in the frequency range from 1.16 THz to 1.634 THz, the Y component of the transmitted wave is ahead of the X component hence the transmitted wave is LHCP, whereas in the frequency range of 3.935 THz to 5.29 THz, the Y component of the transmitted wave lags the X component, hence the transmitted wave is RHCP. In addition, the proposed unit cell behaves equally well for the incident Y polarized wave resulting in RHCP and LHCP for the two frequency bands.

Total transmission in the X direction can be computed as $T_{all} = \left|t_{yx}\right|^2 + \left|t_{xx}\right|^2$ [46]. Figure 5 shows the axial ratio for the incident X polarized and Y polarized waves along with total transmission. For the sake of simplicity, total transmission in the X direction is only shown in Figure 5. A similar tendency is observed for transmission in the Y direction. It is clear from Figure 5 that the proposed structure has an axial ratio of 3 dB from 1.16 THz to 1.634 THz and 3.935 THz to 5.29 THz for both X polarized and Y polarized incident waves. Moreover, reasonable energy transfer (−1 to −5 dB) is observed in the dual-band except in the frequency range 5 THz to 5.29 THz. In fact, there is good energy transfer from 1 THz to 5 THz but the transmitted wave from frequency range 1.634 THz to 3.935 THz is not circularly polarized wave because the axial ratio is much larger than 3 dB.

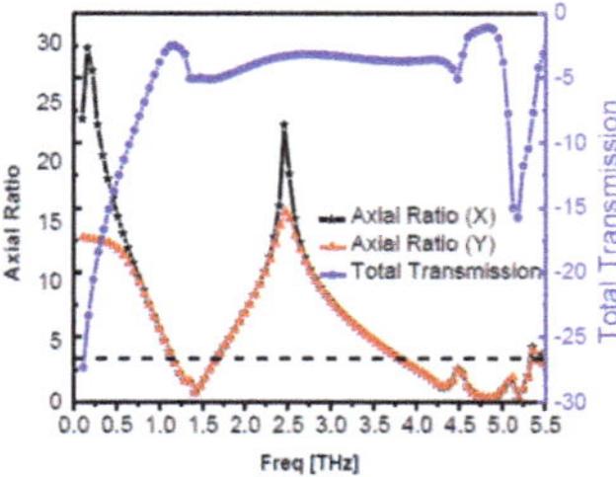

Figure 5. Axial ratio for the transmitted wave with incident X and Y polarized wave and total transmission with incident X polarization.

5. Physical Explanation and Equivalent Circuit

Since the unit cell is based on an anisotropic structure having dual diagonal symmetry, the incident X polarized wave will generate transmitted X and Y polarized wave components and the incident Y polarized wave will generate transmitted X and Y components. To explain the physical phenomenon behind the proposed LP-to-CP, we considered the surface current vectors within two frequency bands; let these be f_1 and f_2: f_1 = 1.398 THz and f_2 = 4.82 THz. Figure 6 shows the surface current distribution with the incident X polarized wave at the output surface of the proposed converter for t = 0, T/4, T/2, 3T/4 at f_1. The orientation of the electric field vectors shows that with every T/4 cycle, it rotates by 90°. Further, it can be seen that the surface current at f_1 is concentrated in the inner tri-square conducting patches with an anti-clockwise rotation. Thus, the transmitted wave is LHCP at f_1.

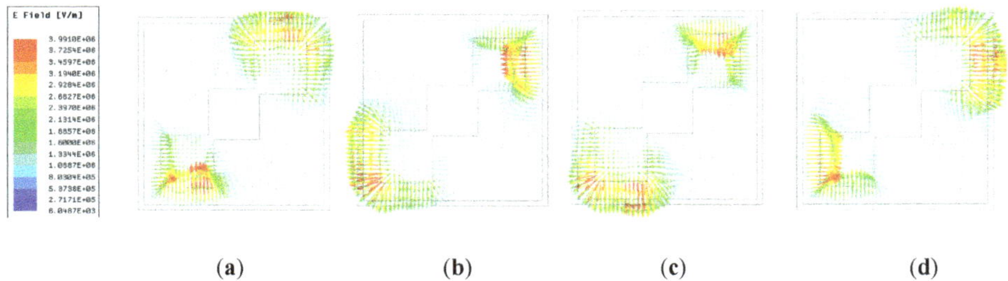

(a)   (b)   (c)   (d)

Figure 6. Surface current distribution of the proposed LP-to-CP converter at 1.398 THz at (**a**) t = 0, (**b**) t = T/4 (**c**) t = T/2 (**d**) t = 3T/4.

Figure 7 shows the surface current vectors at the output surface at f_2. It can be clearly seen that with every quarter cycle, surface currents are rotated 90° in a clockwise rotation. Unlike in Figure 6, this time surface current vectors are concentrated in the outer square ring. The opposite direction of rotation for surface current vectors in time cycle T validates our proposed opposite handedness of circular polarization for the same structure at two different frequencies.

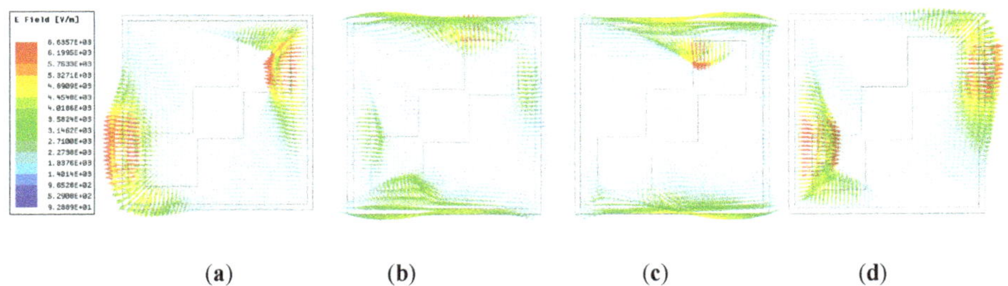

(a)   (b)   (c)   (d)

Figure 7. Surface current distribution of the proposed LP-to-CP converter at 4.82 THz at (**a**) t = 0 (**b**) t = T/4 (**c**) t = T/2 (**d**) t = 3T/4.

Figures 8 and 9 indicate the response of the proposed structure to the incident X polarized electromagnetic field. Figure 8a,b show electric field distribution at 1.398 THz and 4.82 THz, respectively. It is clear from Figure 8a that the electric field concentrates on the outer two conducting patches of the tri-square patch with a minor contribution from corners of an outer square ring, whereas for 4.82 THz, the electric field is concentrated on the whole tri-squares patch and outer square ring. This multi-resonance structure validates the dual-wide-band performance of polarization conversion.

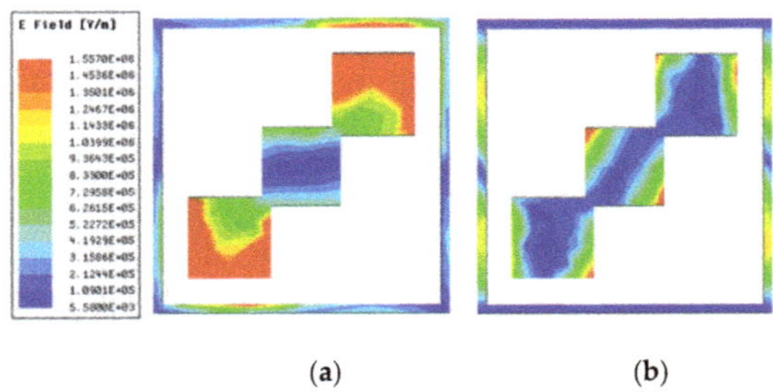

Figure 8. Electric field strength at (a) $f_1 = 1.398$ THz, (b) $f_2 = 4.82$ THz.

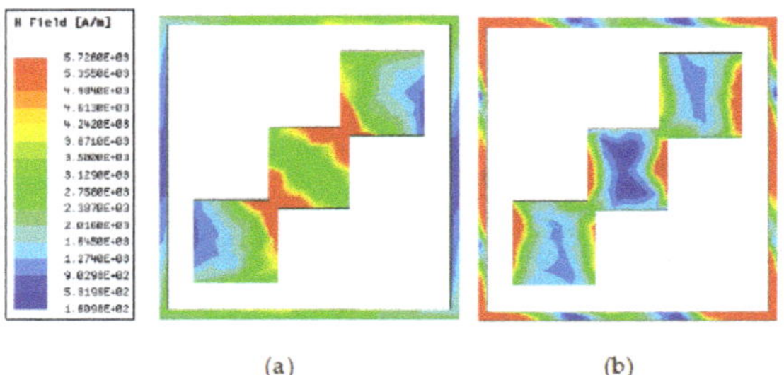

Figure 9. Magnetic field strength at (a) $f_1 = 1.398$ THz, (b) $f_2 = 4.82$ THz.

Figure 9a shows the magnetic field strength at $f_1 = 1.398$ THz. It shows that the magnetic field is concentrated in the intersected corners of the tri-squares conductor patch. Figure 9b shows that for the second frequency band, at f_2, the magnetic field is predominantly attributed to the outer square ring and vertical sides of the squares in the tri-square patch.

Figure 10a,b show the equivalent circuit for the unit cell of the proposed dual-wide-band LP-to-CP converter with incident X and Y polarizations, respectively. With incident X polarizations, upper and lower arms interact with the incident waves, whereas the left and right arms of the outer square ring will have no interaction. Thus, inductors L_1 representing the outer square ring will appear as shown in Figure 10a. C_1 represents the value of capacitance induced between the corner of the square ring and the diagonal conducting patch with incident X polarization. L_2 shows the combined inductive effect of the two outer squares (area: $C \times C$) and inner square (area: $C_1 \times C_1$). It is interesting to note that due to the discontinuity between the three square inductors in a diagonal position, there will also be a capacitive effect, but the overall effect for three inductors will be inductive. Thus, it is represented as L_2. Z_o and Z_d represent transmission line models for free space layers and the substrate. In the lower frequency band of operation, the impedance corresponding to L_1, Z_L_1 will be lower compared to its value in a higher frequency band of operation. Similarly, impedance corresponding to C_1, Z_C_1 will be large in the first frequency band while it will become much smaller in the second frequency band.

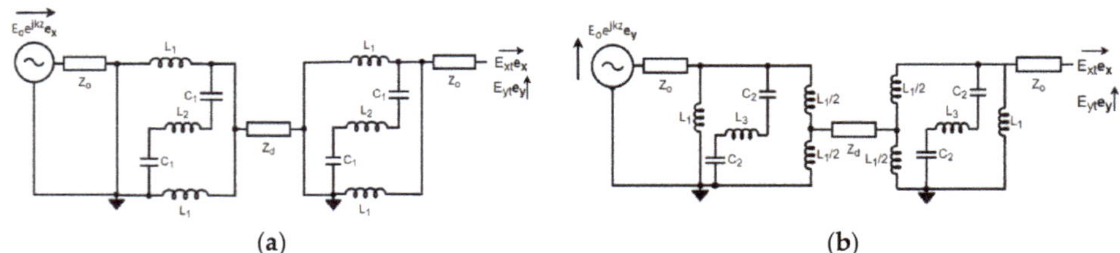

Figure 10. Equivalent circuit for the proposed converter under incident (a) X polarization, (b) Y polarization.

Now it will be explained how dual polarizations exist within two bands. For this, assume Z_L_1 (f_1) and Z_L_1 (f_2) represent impedances corresponding to L_1 at f_1 and f_2, respectively. Similarly, Z_C_1 (f_1) and Z_C_1 (f_2) represent impedances corresponding to C_1 at f_1 and f_2, respectively, and Z_L_2 (f_1) and Z_L_2 (f_2) represent impedances corresponding to L_2 at f_1 and f_2, respectively. f_1 corresponds to any frequency within the first frequency band while f_2 corresponds to any frequency within the second frequency band. Thus, $|Z\_L_1 (f_1)| < |Z\_L_1 (f_2)|$, $|Z\_C_1 (f_1)| >> |Z\_C_1 (f_2)|$, $|Z\_L_2 (f_1)| < |Z\_L_2 (f_2)|$. Overall impedance for the top layer can be calculated as $|Z\_L_1| \parallel |Z\_L_1| \parallel (2 \times |Z\_C_1| + |Z\_L_2|)$. At f_1, $|Z\_C_1(f_1)|$ will be larger than that at f_2, $|Z\_C_1(f_2)|$. Thus, the $2 \times |Z\_C_1| + |Z\_L_2|$ component will be larger. Hence, the overall effect for a parallel combination of large capacitive impedance and small inductive impedance will be small inductive impedance. For a frequency f_2 in higher frequency band, $2 \times |Z\_C_1| + |Z\_L_2|$ will become small and the overall effect will be capacitive. At f_2, the resultant impedance will be a parallel combination of large inductance and small capacitance, which will result in small capacitive effect. Thus, the overall change in impedance behavior at f_2 explains the idea of dual polarization.

6. Performance Analysis

For design considerations, the effect of different dimensions of the unit cell on the performance of the dual-band polarization converter was analyzed. Figure 11a shows the plot of the axial ratio for the proposed converter for different values of p keeping all the other parameters constant. It is observed that with the increase in p from 49.9 μm to 51.9 μm, the lower end of the first frequency band remains almost constant while higher-end shifts towards the right cause the bandwidth of the first band to increase. In the second band, increasing p has greater impact: it shifts the lower frequency towards the left while the higher-end remains almost stable with some exceptions. Thus, periodicity p has an impact on both conversion bands. This can be explained as follows: increasing p increases the length of the outer square ring on the metasurface, since this square ring contributes to the electric field strength at f_1 less than that at f_2 (as seen in Figure 8). Thus, variation in the second conversion band is found to be larger than for the first conversion band. Figure 11b shows the plot of the axial ratio for different values of c from 13.2 μm to 14.2 μm. It is clear that with the increase in c, the first frequency band shifts towards the left, which decreases operational bandwidth. A similar tendency is observed for the second frequency band due to the larger variation per wavelength as compared to the first frequency band. This change is supported by Figure 8, in which both conversion bands' operation depends upon corner square-conducting patches.

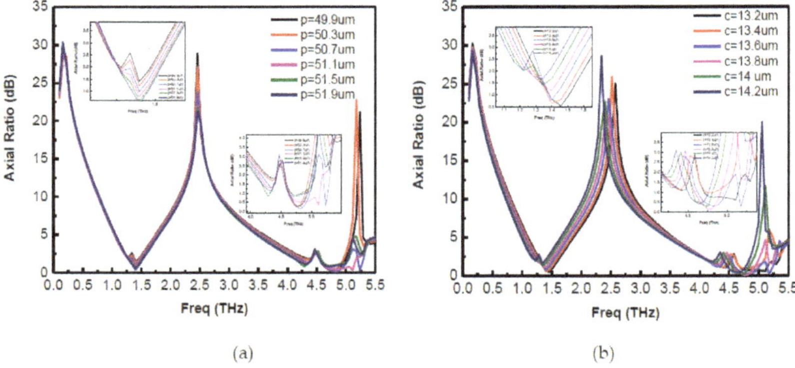

Figure 11. Effect of (a) p and (b) c on the performance of dual-band LP-to-CP converter.

The effect of substrate thickness d and inner square dimension c_1 is analyzed as shown in Figure 12a,b, respectively. Figure 12a shows that as d is increased from 3.2 μm to 4.2 μm, the performance of dual-band LP-to-CP deteriorates in terms of axial ratio. A slight decrease in bandwidth is observed as the higher frequency end of the first band moves towards the left. This variation in second frequency band seems to be abrupt due to higher sensitivity at high frequency, although performance remains more or less stable (within 3 dB), except at $d = 4.2$ μm. This phenomenon can be explained quantitatively as follows: the incident electric field on the top metasurface can be divided into two parts: the reflected wave to the air and the transmitted wave inside the substrate. Assuming the conductor thickness to be negligible, the transmitted wave travels inside the substrate and upon striking the substrate to the ground interface, is partially reflected back to the substrate, whereas the remainder of the portion is transmitted into the air. The portion of electromagnetic wave which was reflected back to the substrate travels back to the top surface to the substrate interface and, upon striking that interface, is again divided into two portions. One portion is reflected back to the substrate, while the other portion is transmitted into the air. The reflected wave traveling inside the substrate experiences the phase delays and, upon striking the substrate to ground interface, some portion of this wave reflects back to the substrate, while some part is transmitted into the air. These multiple transmitted waves interfere with each other constructively and destructively producing two polarized waves: one X polarized and the other Y polarized. These waves generate circularly polarized waves when they have comparable magnitudes and differences in phase angles around 90°. Varying the substrate thickness changes the phase angles and hence affects the axial ratio for the proposed converter. Figure 12b shows the effect of $c1$ on the performance of dual-band LP-to-CP converter. It shows that the first frequency band remains almost stable with a change in $c1$, whereas the second frequency band is sensitive towards $c1$. Although its performance remains within 3 dB for 13.76 to 14.06 um, the value of $c1$ affects the second frequency band. The same is obvious from the physical mechanisms discussed earlier and as shown in Figure 8, where it is clear that the central conducting patch contributes to the higher frequency band and does not significantly impact the lower frequency band.

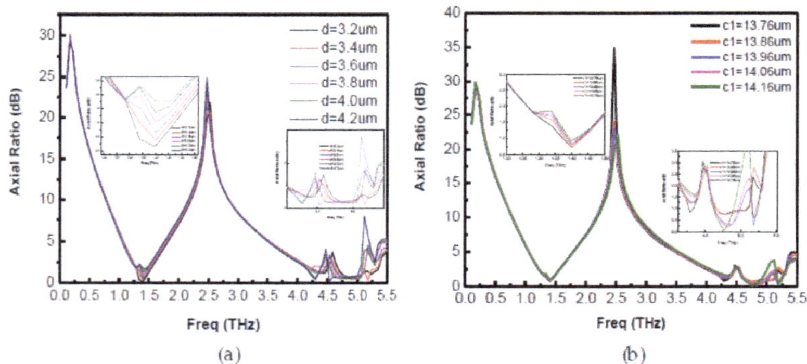

Figure 12. Effect of (**a**) d and (**b**) c_1 on the performance of dual-band LP-to-CP converter.

7. Conclusions

In summary, a bi-layered metasurface-based THz dual-wide-band LP-to-CP converter based on transmission characteristics is proposed. For incident linear X or Y polarized waves, it transmits dual polarized dual-band circularly polarized waves. It works over 1.16 THz to 1.634 THz (34% fractional bandwidth) and 3.935 THz to 5.29 THz (29% fractional bandwidth) within two bands for an axial ratio of 3 dB. Considerable energy transfer is observed in dual bands except for the band 5 THz to 5.29 THz. Close agreement between HFSS (FEM) and CST (FDTD) results ensures the validity and reliability of the proposed design. Physical mechanisms, electrical circuit representation and quantitative explanation of the working principle behind the dual-wide-band LP-to-CP conversion justify the designed structure. Such a simple and dual-wide-band, dual polarized LP-to-CP converter can be used in potential applications in THz wireless communication, including satellite communication, ground segment, imaging, spectrometry, antenna technology, and many other THz manipulating devices.

Author Contributions: Writing-Original Draft Preparation, A.K.F.; Writing-Review and editing, C.R. and K.C.; methodology, A.K.F.; software, A.K.F.; validation, K.C.; and C.R.; supervision, C.R., project administration, C.R.

Funding: This research was funded by the National Natural Science Foundation of China, grant number 61831001, the High-Level Talent Introduction Project of Beihang University, grant number ZG216S1878, and the Youth-Top-Talent Support Project of Beihang University, grant number ZG226S1821.

Acknowledgments: Authors convey their thanks to Prof. Jinjun Feng in the Vacuum Electronics National Laboratory, Beijing Vacuum Electronics Research Institute, China for his sincere supports for the CST software.

Conflicts of Interest: The authors declare no conflict of interest.

References

1. Jepsen, P.U.; Cooke, D.G.; Koch, M. Terahertz spectroscopy and imaging–Modern techniques and applications. *Laser Photonics Rev.* **2011**, *5*, 124–166. [CrossRef]
2. Katletz, S.; Pfleger, M.; Pühringer, H.; Mikulics, M.; Vieweg, N.; Peters, O.; Scherger, B.; Scheller, M.; Koch, M.; Wiesauer, K. Polarization sensitive terahertz imaging: Detection of birefringence and optical axis. *Opt. Express* **2012**, *20*, 23025–23035. [CrossRef] [PubMed]
3. Tonouchi, M. Cutting-edge terahertz technology. *Nat. Photonics* **2007**, *1*, 97. [CrossRef]
4. Song, Z.; Gao, Z.; Zhang, Y.; Zhang, B. Terahertz transparency of optically opaque metallic films. *EPL Europhys. Lett.* **2014**, *106*, 27005. [CrossRef]
5. Fang, Z.H.; Chen, H.; An, D.; Luo, C.R.; Zhao, X.P. Manipulation of visible-light polarization with dendritic cell-cluster metasurfaces. *Sci. Rep.* **2018**, *8*, 9696. [CrossRef] [PubMed]
6. Cui, J.; Huang, C.; Pan, W.; Pu, M.; Guo, Y.; Luo, X. Dynamical manipulation of electromagnetic polarization using anisotropic meta-mirror. *Sci. Rep.* **2016**, *6*, 30771. [CrossRef]
7. Cheng, Y.; Gong, R.; Wu, L. Ultra-broadband linear polarization conversion via diode-like asymmetric transmission with composite metamaterial for terahertz waves. *Plasmonics* **2017**, *12*, 1113–1120. [CrossRef]
8. Wang, S.Y.; Liu, W.; Geyi, W. Dual-band transmission polarization converter based on planar-dipole pair frequency selective surface. *Sci. Rep.* **2018**, *8*, 3791. [CrossRef]
9. Kaveev, A.K.; Kropotov, G.I.; Tsypishka, D.I.; Tzibizov, I.A.; Vinerov, I.A.; Kaveeva, E.G. Tunable wavelength terahertz polarization converter based on quartz waveplates. *Appl. Opt.* **2014**, *53*, 5410–5415. [CrossRef]
10. Zi, J.; Xu, Q.; Wang, Q.; Tian, C.; Li, Y.; Zhang, X.; Han, J.; Zhang, W. Terahertz polarization converter based on all-dielectric high birefringence metamaterial with elliptical air holes. *Opt. Commun.* **2018**, *416*, 130–136. [CrossRef]
11. Strikwerda, A.C.; Fan, K.; Tao, H.; Pilon, D.V.; Zhang, X.; Averitt, R.D. Comparison of birefringent electric split-ring resonator and meanderline structures as quarter-wave plates at terahertz frequencies. *Opt. Express* **2009**, *17*, 136–149. [CrossRef] [PubMed]
12. Lin, X.W.; Wu, J.B.; Hu, W.; Zheng, Z.G.; Wu, Z.J.; Zhu, G.; Xu, F.; Jin, B.B.; Lu, Y.Q. Self-polarizing terahertz liquid crystal phase shifter. *Aip Adv.* **2011**, *1*, 032133. [CrossRef]
13. Hsieh, C.F.; Lai, Y.C.; Pan, R.P.; Pan, C.L. Polarizing terahertz waves with nematic liquid crystals. *Opt. Lett.* **2008**, *33*, 1174–1176. [CrossRef]
14. Wang, L.; Ge, S.; Hu, W.; Nakajima, M.; Lu, Y. Tunable reflective liquid crystal terahertz waveplates. *Opt. Mater. Express* **2017**, *7*, 2023–2029. [CrossRef]
15. Reid, M.; Fedosejevs, R. Terahertz birefringence and attenuation properties of wood and paper. *Appl. Opt.* **2006**, *45*, 2766–2772. [CrossRef]
16. Scherger, B.; Scheller, M.; Vieweg, N.; Cundiff, S.T.; Koch, M. Paper terahertz wave plates. *Opt. Express* **2011**, *19*, 24884–24889. [CrossRef] [PubMed]
17. Huang, Y.; Yao, Z.; Hu, F.; Liu, C.; Yu, L.; Jin, Y.; Xu, X. Tunable circular polarization conversion and asymmetric transmission of planar chiral graphene-metamaterial in terahertz region. *Carbon* **2017**, *119*, 305–313. [CrossRef]
18. Kanda, N.; Konishi, K.; Kuwata-Gonokami, M. Terahertz wave polarization rotation with double layered metal grating of complimentary chiral patterns. *Opt. Express* **2007**, *15*, 11117–11125. [CrossRef]
19. Kenanakis, G.; Zhao, R.; Stavrinidis, A.; Konstantinidis, G.; Katsarakis, N.; Kafesaki, M.; Soukoulis, C.M.; Economou, E.N. Flexible chiral metamaterials in the terahertz regime: A comparative study of various designs. *Opt. Mater. Express* **2012**, *2*, 1702–1712. [CrossRef]

20. Wu, P.C.; Zhu, W.; Shen, Z.X.; Chong, P.H.J.; Ser, W.; Tsai, D.P.; Liu, A.Q. Broadband Wide-Angle Multifunctional Polarization Converter via Liquid-Metal-Based Metasurface. *Adv. Opt. Mater.* **2017**, *5*, 1600938. [CrossRef]

21. Yan, L.; Zhu, W.; Karim, M.F.; Cai, H.; Gu, A.Y.; Shen, Z.; Chong, P.H.J.; Tsai, D.P.; Kwong, D.L.; Qiu, C.W.; et al. Arbitrary and Independent Polarization Control in Situ via a Single Metasurface. *Adv. Opt. Mater.* **2018**, *6*, 1800728. [CrossRef]

22. Chen, W.T.; Török, P.; Foreman, M.R.; Liao, C.Y.; Tsai, W.Y.; Wu, P.R.; Tsai, D.P. Integrated plasmonic metasurfaces for spectropolarimetry. *Nanotechnology* **2016**, *27*, 224002. [CrossRef] [PubMed]

23. Jiang, Y.; Wang, L.; Wang, J.; Akwuruoha, C.N.; Cao, W. Ultra-wideband high-efficiency reflective linear-to-circular polarization converter based on metasurface at terahertz frequencies. *Opt. Express* **2017**, *25*, 27616–27623. [CrossRef] [PubMed]

24. Lin, B.Q.; Guo, J.; Wang, Y.; Wang, Z.; Huang, B.; Liu, X. A Wide-Angle and Wide-Band Circular Polarizer Using a Bi-Layer Metasurface. *Prog. Electromagn. Res.* **2018**, *161*, 125–133. [CrossRef]

25. Zang, X.F.; Liu, S.J.; Gong, H.H.; Wang, Y.; Zhu, Y.M. Dual-band superposition induced broadband terahertz linear-to-circular polarization converter. *JOSA B* **2018**, *35*, 950–957. [CrossRef]

26. Gao, X.; Han, X.; Cao, W.P.; Li, H.O.; Ma, H.F.; Cui, T.J. Ultrawideband and high-efficiency linear polarization converter based on double V-shaped metasurface. *IEEE Trans. Antennas Propag.* **2015**, *63*, 3522–3530. [CrossRef]

27. Ji-Bao, Y.; Hua, M.; Jia-Fu, W.; Ming-De, F.; Yong-Feng, L.; Shao-Bo, Q. High-efficiency ultra-wideband polarization conversion metasurfaces based on split elliptical ring resonators. *Acta Phys. Sin.* **2015**, *64*. [CrossRef]

28. Hadad, Y.; Sounas, D.L.; Alu, A. Space-time gradient metasurfaces. *Phys. Rev. B* **2015**, *92*, 100304. [CrossRef]

29. Taravati, S.; Kishk, A.A. Advanced wave engineering via obliquely illuminated space-time-modulated slab. *IEEE Trans. Antennas. Propag.* **2018**, *67*, 270–281. [CrossRef]

30. Taravati, S.; Kishk, A.A. Dynamic modulation yields one-way beam splitting. *Phys. Rev. B* **2019**, *99*, 075101. [CrossRef]

31. Shi, Y.; Han, S.; Fan, S. Optical circulation and isolation based on indirect photonic transitions of guided resonance modes. *ACS Photonics* **2017**, *4*, 1639–1645. [CrossRef]

32. Zang, J.W.; Correas-Serrano, D.; Do, J.T.S.; Liu, X. Alvarez-Melcon, A.; Gomez-Diaz, J.S. Nonreciprocal wavefront engineering with time-modulated gradient metasurfaces. *Phys. Rev. Appl.* **2019**, *11*, 054054. [CrossRef]

33. Zhao, J.X.; Xiao, B.X.; Huang, X.J.; Yang, H.L. Multiple-band reflective polarization converter based on complementary L-shaped metamaterial. *Microw. Opt. Technol. Lett.* **2015**, *57*, 978–983. [CrossRef]

34. Liu, Y.; Xia, S.; Shi, H.; Zhang, A.; Xu, Z. Dual-band and high-efficiency polarization converter based on metasurfaces at microwave frequencies. *Appl. Phys. B* **2016**, *122*, 178. [CrossRef]

35. Huang, X.; Yang, H.; Zhang, D.; Luo, Y. Ultrathin Dual-band Metasurface Polarization Converter. *IEEE Trans. Antennas. Propag.* **2019**, *67*, 4636–4641. [CrossRef]

36. Zeng, Q.; Ren, W.; Zhao, H.; Xue, Z.; Li, W. Dual-band transmission-type circular polariser based on frequency selective surfaces. *IET Microw. Antennas Propag.* **2018**, *13*, 216–222. [CrossRef]

37. Naseri, P.; Matos, S.A.; Costa, J.R.; Fernandes, C.A.; Fonseca, N.J. Dual-Band Dual-Linear-to-Circular Polarization Converter in Transmission Mode Application to K/Ka-Band Satellite Communications. *IEEE Trans. Antennas. Propag.* **2018**, *66*, 7128–7137. [CrossRef]

38. Wang, H.B.; Cheng, Y.J. Single-Layer Dual-band Linear-to-Circular Polarization Converter with Wide Axial Ratio Bandwidth and Different Polarization Modes. *IEEE Trans. Antennas Propag.* **2019**, *6*, 4296–4301. [CrossRef]

39. Youn, Y.; Hong, W. Planar dual-band linear to circular polarization converter using radial-shape multi-layer FSS. In Proceedings of the 2018 IEEE International Symposium on Antennas and Propagation & USNC/URSI National Radio Science Meeting, Boston, MA, USA, 8–13 July 2018.

40. Yang, S.; Jiang, Y.; Wang, J.; Zhao, H. Dual-Ultrawideband Linear-to-Circular Converter with Double Rotation Direction in Terahertz Frequency. In Proceedings of the 2018 Cross Strait Quad-Regional Radio Science and Wireless Technology Conference (CSQRWC), Xuzhou, China, 21–24 July 2018.

41. Li, Y.; Zhang, J.; Qu, S.; Wang, J.; Zheng, L.; Pang, Y.; Xu, Z.; Zhang, A. Achieving wide-band linear-to-circular polarization conversion using ultra-thin bi-layered metasurfaces. *J. Appl. Phys.* **2015**, *117*, 044501. [CrossRef]

42. Perez-Palomino, G.; Page, J.E.; Arrebola, M.; Encinar, J.A. A design technique based on equivalent circuit and coupler theory for broadband linear to circular polarization converters in reflection or transmission mode. *IEEE Trans. Antennas Propag.* **2018**, *66*, 2428–2438. [CrossRef]

43. Akgol, O.; Unal, E.; Altintas, O.; Karaaslan, M.; Karadag, F.; Sabah, C. Design of metasurface polarization converter from linearly polarized signal to circularly polarized signal. *Optik* **2018**, *161*, 12–19. [CrossRef]

44. Akgol, O.; Altintas, O.; Unal, E.; Karaaslan, M.; Karadag, F. Linear to left-and right-hand circular polarization conversion by using a metasurface structure. *Int. J. Microw. Wirel. Technol.* **2018**, *10*, 133–138. [CrossRef]

45. Altintas, O.; Unal, E.; Akgol, O.; Karaaslan, M.; Karadag, F.; Sabah, C. Design of a wide band metasurface as a linear to circular polarization converter. *Mod. Phys. Lett. B* **2017**, *31*, 1750274. [CrossRef]

46. Lin, B.; Guo, J.; Ma, Y.; Wu, W.; Duan, X.; Wang, Z.; Li, Y. Design of a wideband transmissive linear-to-circular polarization converter based on a metasurface. *Appl. Phys. A* **2018**, *124*, 715. [CrossRef]

Article

Two-Dimensional Imaging of Permittivity Distribution by an Activated Meta-Structure with a Functional Scanning Defect

Go Itami [1,2,*], Osamu Sakai [1,2] and Yoshinori Harada [3]

[1] Department of Electronic Systems Engineering, The University of Shiga Prefecture, Shiga 522-8533, Japan; sakai.o@e.usp.ac.jp
[2] Electric Science and Engineering, Kyoto University, Kyoto 615-8510, Japan
[3] Graduate School of Medicine, Kyoto Prefectural University of Medicine, Kyoto 602-8566, Japan; yoharada@koto.kpu-m.ac.jp
* Correspondence: ot68gitami@ec.usp.ac.jp

Received: 31 January 2019; Accepted: 18 February 2019; Published: 20 February 2019

Abstract: A novel 2D imaging method for permittivity imaging using a meta-structure with a functional scanning defect is proposed, working in the millimeter wave-range. The meta-structure we used here is composed of a perforated metal plate with subwavelength-holes and a needle-like conductor that can scan two-dimensionally just beneath the plate. The metal plate, which is referred to as a metal hole array (MHA) in this study, is known as a structure supporting propagation of spoof surface plasmon polaritons (SSPPs). High-frequency waves with frequencies higher than microwaves, including SSPPs, have the potential to detect signals from inner parts embedded beneath solid surfaces such as living cells or organs under the skin, without physical invasion, because of the larger skin depth penetration of millimeter wave-bands than optical wave-bands. Focused on activated SSPPs, the localized distortion of SSPP modes on an MHA is used in the proposed method to scan the electromagnetic properties of the MHA with a needle-like conductor (conductive probe), which is a kind of active defect-initiator. To show the validity of the proposed method, electromagnetic analyses of the localized distortions of wave fields were performed, and one- and two-dimensional imaging experiments were conducted with the aim of detecting both conductive and dielectric samples. The analytical results confirmed the localized distortion of the electric field distribution of SSPP modes and also indicated that the proposed method has scanning ability. In experimental studies, the detection of conductive and dielectric samples was successful, where the detected dielectrics contained pseudo-biological materials, with an accuracy on the order of millimeters. Finally, a biomedical diagnosis in the case of a rat lung is demonstrated by using the experimental system. These results indicate that the proposed method may be usable for non-invasive and low-risk biomedical diagnosis.

Keywords: spoof surface plasmon polariton (SSPP); metal hole arrays (MHA); electromagnetic distortion; two-dimensional imaging

---

1. Introduction

Electromagnetic waves have huge potential for various applications such as wireless communication, micro processing, and sensing, and as a result, they will become indispensable in our daily lives. In recent years, millimeter waves or terahertz waves, with wavelengths shorter than microwaves and longer than infrared, have been attracting considerable attention. One of the features of these waves is their combination of high resolution and non-destructiveness. Therefore, a number of imaging methods that use these waves have been reported [1–3]. However, since these waves generally propagate in straight lines and attenuate immediately in lossy media, such methods frequently encounter technical issues due to the deterioration of information-carrying detection signals and their physical separation due to multipath propagation. From a microscopic point of view, an electromagnetic wave includes two components: electric and magnetic fields with their spatial distributions varying in time. This is significantly different from other waves, like sonic waves, which can be interpreted in terms of single-variable wave propagation.

When we impose specific boundary conditions on electromagnetic waves, there is a wave component with a wavenumber vector that does not propagate spatially and concentrates around the boundaries or interfaces with exponential attenuation. This phenomenon corresponds to an evanescent wave, and in classical cases it has only been observed in the optical region [4]. In a microscopic view, when the evanescent waves get close to a conductive plate, oscillations are excited in the free electrons in the plate surface. This phenomenon is called a surface plasmon polariton (SPP) or simply, a surface wave [5–8], and it has been used in applications such as biosensors, chemical sensors and field enhancement in spectroscopy [9–11]. As described above, although an SPP can be excited only in the optical region, a similar phenomenon, with the concentration and oscillation of electric fields on the surface of a conductive plate, has also been confirmed at lower frequencies than optical bands. These are called spoof surface plasmon polaritons (SSPP) [12–16], and are broadly considered to be a unique phenomenon involving a metamaterial. One of the features of SSPPs that is not observed in other propagating modes, is the concentration of its electric field at interfaces, so that it has the potential to give us significant information about objects near such surfaces without complex signal processing. The standard example of an SSPP structure is a metal hole array (MHA), which is a conductive plate with periodic holes [17–20]. An MHA is considered to be a macroscopic effective medium in terms of controlling relative permittivity by the structural parameters of its holes. Since its structure has only a few parameters, an MHA is expected to serve as a ready-to-use functional and practical medium as well as a readily available sample for experiments [21]. Therefore, many studies of SSPPs on MHAs have been reported, although so far, they have focused on static properties as SSPP applications.

In this study, a two-dimensional imaging method using an MHA as a dynamic functional material is proposed for permittivity imaging, mainly aimed at a non-invasive medical diagnostic. Specifically, the proposed method uses the intentional formation of electromagnetic defects in the uniform electric distributions of SSPPs on an MHA, where the defect is formed by scanning a metallic needle across the surface of the MHA. Viewed another way, the proposed method is also a non-destructive diagnostic method used as a dynamic dielectric sensor for an MHA, which is sensitive to the side opposite the needle-scanned surface. Compared to other non-destructive diagnosis tools such as confocal microscopy and near-field optical microscopy, an advantage of the proposed method is that the electromagnetic properties of subsurface tissues can be measured because surface waves in millimeter or terahertz wave bands have deeper skin penetration depths than those in optical frequency bands [22–24]. At present, the imaging of inner issues is typically achieved using a destructive method such as CT (Computer Tomography) or some other high-cost technique such as MRI (Magnetic Resonance Imaging). Actually, conventional imaging methods face difficulties arising from these issues [25–27]. Biomedical imaging based on the electromagnetic properties of biomedical tissues is considered to be an effective method because the electromagnetic properties of a malignant tumor and normal breast tissue are different [28].

In Section 2, the excitation mechanism of SSPPs on an MHA is described. In Section 3, the numerical analysis of the electromagnetic properties of an MHA, such as transmittance and reflectance, and the localized distortion of the electric fields of the SSPP mode around the MHA with a conductive probe (a needle-like conductor) is demonstrated. In Section 4, we report one- and two-dimensional imaging experiments including the experimental demonstration of biomedical diagnosis in the case of a rat lung that apply the localized distortion mechanism, with the use of an MHA targeting conductive, dielectric, and biomedical samples.

2. The Basis of Spoof Surface Plasmon Generation

The SSPP phenomenon was discovered by Pendry et al. who also derived the theory of its generation. Here, SSPP is used for imaging by detecting the local deformation of its propagation modes. Therefore, in order to explain the proposed imaging method, we begin with the theoretical derivation of SSPP generation as its basis. First of all, the mechanism of SSPP generation on an MHA is explained. In order to describe SSPP generation on an MHA, the electromagnetic model of the process is introduced. In this study, the MHA has holes whose size is the same order of magnitude as the wavelengths of incident waves. The model is shown in Figure 1. In this model, the MHA is treated as a uniform, effective medium with a frequency-based response to incident waves. As shown in Figure 1, we consider that with the incidence of optimized millimeter waves, electric fields concentrate in and around an MHA that has two-dimensional waveguide arrays (period d, cross section $p \times q$, depth w,) extended to infinity. It is assumed that the incident waves are TM-polarized optimal waves. Note that the electric field-vector of the wave has only x- and z- components, and the magnetic field-vector has only y-component in the situation. In the derivations, the coupling phenomenon between the dispersion relation of the incident waves and the waveguide modes is considered. That is, the electric fields are considered to be concentrated on the surface, and this distribution forms longitudinal waves. Thus, the SSPP modes are generated on the MHA.

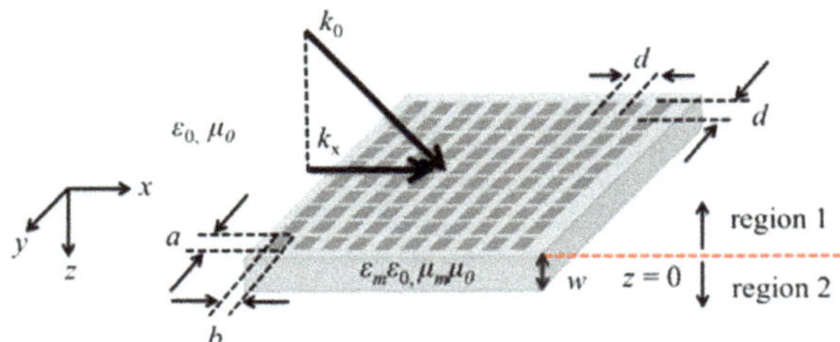

Figure 1. The theoretical model of the metal hole array (MHA) with TM incident waves. Region 1 is the space above the MHA. Region 2 is the space in the holes of the MHA.

Before beginning the derivation of SSPP generation, the definitions of the electric and magnetic properties in two regions are introduced. In Figure 1, Region 1 indicates the space on the upper side of the MHA, and Region 2 indicates the space inside the holes of the MHA. Since Region 1 is filled with vacuum, the permittivity and the permeability in Region 1 are ε_0 and μ_0 respectively. On the other hand, as described above, since the MHA is presumed to be a macroscopically uniform medium, the MHA has to have an effective relative permittivity of ε_m and a relative permeability μ_m that is frequency dependent [12]. When electromagnetic waves whose frequency is below a cut-off frequency are introduced to an MHA, waves are attenuated with fundamental mode [29]. Therefore, the electric fields of incident waves are expressed as below.

$$E_x = E_0 \sin\left(\frac{\pi y}{a}\right) \tag{1}$$

Note that in this manuscript, time oscillation terms are omitted for convenience of formulation if necessary. Here, E_0 is a constant value, and expression (1) indicates that $E_x = 0$ at the edges of the holes. Therefore, the x-component and z-component of an incident wavevector k_x and k_z can be expressed as

$$k_x = \frac{\pi}{a} \tag{2}$$

$$k_z = \sqrt{\left(\frac{\pi}{a}\right)^2 - \omega^2 \varepsilon_h \mu_h \varepsilon_0 \mu_0} \tag{3}$$

where ε_h and μ_h are respectively, the relative permittivity and relative permeability in the holes of the MHA. At this point, remembering the precondition that the MHA is a macroscopic medium and has an effective relative permittivity, for example, when focused on a unit cell of the MHA (shown in Figure 2), the macroscopic wave vector only has a component in the z-direction k_z' because the waveguide mode occurs only in the holes and provides no propagation in the x- and y- directions. From the above discussions, k_z' can be expressed as below.

$$k'_z = k_0\sqrt{\varepsilon_x\mu_y} = k_0\sqrt{\varepsilon_m\mu_m},\ k_0 = \frac{\omega}{c_0} \tag{4}$$

where c_0 is the velocity of light in a vacuum. The wave number of k_z' and that of k_z should be essentially the same value, because the unit cell in Figure 2 has one waveguide and there is no propagation in the conductive medium of the MHA. Therefore, the equation below is obtained by using the expression $k_z = k_z'$.

$$\frac{\omega}{c_0}\sqrt{\varepsilon_m\mu_m} = \sqrt{\left(\frac{\pi}{a}\right)^2 - \omega^2\varepsilon_h\mu_h\varepsilon_0\mu_0} \tag{5}$$

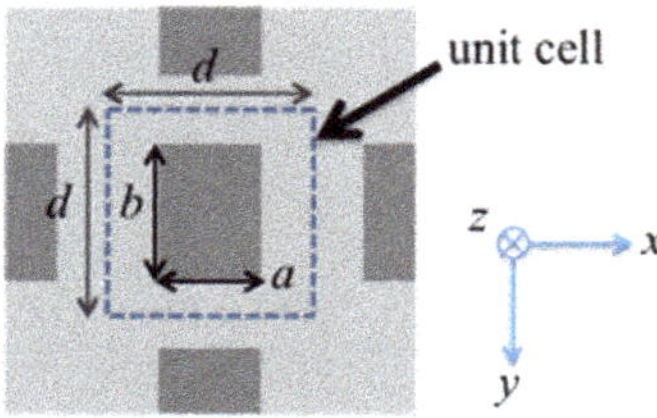

Figure 2. The unit cell of the MHA.

On the other hand, in the matching procedure at $z = 0$, the instantaneous flow of energy across the surface has to be considered by treating the unit cell as both a microscopic and a macroscopic medium. With the average electric fields of the unit cell regarded as a macroscopic medium

$$\overline{E_0} = E_0\frac{a}{d^2}\int_0^b \sin\left(\frac{\pi y}{a}\right)dy = \frac{2ab}{\pi d^2}E_0 \tag{6}$$

$$(E \times H)_{z,\ mic} = \frac{-k_z E_0^2}{\omega\mu_h\mu_0}\frac{a}{d^2}\int_0^b \sin^2\left(\frac{\pi y}{a}\right)dy = \frac{-k_z}{\omega\mu_h\mu_0}E_0^2\frac{ab}{2d^2} \tag{7}$$

$$(E \times H)_{z,\ mac} = \frac{-k_z\overline{E_0}^2}{\omega\mu_m\mu_0} \tag{8}$$

Since the two expressions (7) and (8) for the instantaneous flow of energy across the surface must have the same value, the equation $(E \times H)_{z,\ mic} = (E \times H)_{z,\ mac}$ is always useful. With this equation and expressions (5)–(8), the effective relative permittivity ε_m and relative permeability μ_m can be obtained as below.

$$\mu_m = \mu_h\frac{8ab}{\pi^2 d^2} \tag{9}$$

$$\varepsilon_m = \frac{\pi^2 d^2}{8ab}\left[1 - \left(\frac{\omega_p}{\omega}\right)^2\right],\ \omega_p = \frac{\pi c}{\sqrt{\varepsilon_h\mu_h}a} \tag{10}$$

where ω_p is a cutoff frequency, and ε_h is the relative permittivity and μ_h is the relative permeability in the holes of the MHA. Expression (10) confirms that the effective relative permittivity of the MHA is frequency dependent. Thus, the MHA can be treated as an electromagnetic functional material.

On the other hand, remembering the magnetic field of the incident waves, this can be also solved by a wave equation for the value of H_y. Here, in discussing the generation of SSPPs in the MHA, we start by considering the magnetic field near the MHA. First, the wave equation for H_y is obtained by considering the Maxwell equations.

$$\left(\frac{\partial^2}{\partial x^2} + \frac{\partial^2}{\partial z^2}\right)H_y = -\omega^2\varepsilon_r\mu_r\varepsilon_0\mu_0 H_y \tag{11}$$

where ε_r and μ_r are the relative permittivity and the relative permeability. With the precondition that the TM wave is incident to the MHA, it is considered that if SSPPs occur on the surface of the MHA, they must propagate along the x-direction. Also, since the incident magnetic field has the propagation constants of x and z, the y component of the magnetic field vector H_y can be expressed thus,

$$H_y = h(z)\exp\left[j(k_{\parallel}x - \omega t)\right] \tag{12}$$

where $k_{\parallel}$ is the wavenumber of an SSPP, and $h(z)$ is the amplitude of the magnetic field, depending only on the value of z. Substituting expression (12) into expression (11), the equation for $h(z)$ is obtained. It has the form

$$\frac{\partial^2}{\partial z^2}h(z) = (k_i^2 - \omega^2\varepsilon_r\mu_r\varepsilon_0\mu_0)h(z) \tag{13}$$

where k_i is the wavenumber of the magnetic field. In Region 1, $k_1 = (k_{\parallel}^2 - \omega^2\varepsilon_0\mu_0)^{1/2}$, in Region 2, $k_2 = (-\omega^2\varepsilon_m\mu_m\varepsilon_0\mu_0)^{1/2}$. In Region 2, when regarding the MHA as a macroscopic medium, there is no propagation of the EM waves in x-y directions, and k_1 and k_2 are required to have positive values. Considering the solution of equation (13), as the value of $|z|$ increases, the value of the amplitude $h(z)$ is supposed to decrease exponentially. From these assumptions, the relative equation obtained is:
At $z < 0$

$$h(z) = h_1 \exp(k_1 z) \tag{14}$$

On the other hand, at $z \geq 0$

$$h(z) = h_2 \exp(-k_2 z) \tag{15}$$

When considering the boundary condition of the electromagnetic field and magnetic field between Region 1 and Region 2 (at $z = 0$), the tangent components of the electric field and the magnetic field in Region 1 and Region 2 must have the same value. This condition corresponds to the condition of conventional surface plasmon generation, that is

$$k_1 = -\frac{k_2}{\varepsilon_m}, \; (h_1 = h_2) \tag{16}$$

The above expression is often used to derive the dispersion relation of a surface plasmon. However, expression (16) can also be used to derive the dispersion relation of the SSPP. By squaring expression (16) and using expression (10) and the values of k_1 and k_2, the dispersion relation that is fundamental to comprehending the results in this paper is obtained. That is,

$$k_{\parallel}^2 c_0^2 = \omega^2 + \frac{\omega^4}{\omega_p^2 - \omega^2}\left(\frac{8ab}{\pi^2 d^2}\right)^2 \tag{17}$$

Note that the formula $c_0^2 = 1/\omega^2\varepsilon_0\mu_0$ is used when deriving expression (17). The above expression is called the SSPP dispersion relation. The curve of the dispersion relation is shown in Figure 3. If an angular frequency ω is close to the cutoff frequency ω_p, the wavenumber $k_{\parallel}$ diverges to infinity. In other words, the SSPP modes stop propagating then. On the other hand, as the value of ω approaches zero, the dispersion relation curve comes close to a light line, although they do not cross each other, as shown in Figure 3. Therefore, if an actual dispersion relation corresponds to this curve, SSPP modes are not generated. However, since the MHA is a periodic structure, it has to be treated as a two-dimensional crystal. Therefore, it should be considered that lattice scattering effects are occurring in the MHA. Here, introducing the reciprocal lattice of its structure, the definition of the wavenumber $k_{||}$ can be replaced as

$$k'_{||} = k_{||} - iG_x - jG_y, \quad \left|G_x\right| = \left|G_y\right| = \pi/d \tag{18}$$

where G_x and G_y are the reciprocal lattice vectors in the x-direction and y-direction, respectively. So, the dispersion relation of the MHA in infinite space is also replaced by one with the spatial periodicity shown in Figure 4. Figure 4 confirms that there are intersection points between the actual dispersion relation and a light line. Therefore, SSPPs are indeed generated in the MHA, especially when the angular frequency ω is close to the cutoff frequency ω_p. If the SSPPs are generated on the surface of the incidence side of the MHA, the waves are also generated on the opposite side because the waves transmitted through the waveguides of the MHA also couple with the SSPP mode in the same way as with the incident waves. The important feature of an SSPP mode is the concentration of the electric field on the MHA. Therefore, we focused on a modification of the concentrated electric fields for a novel use as an imaging method.

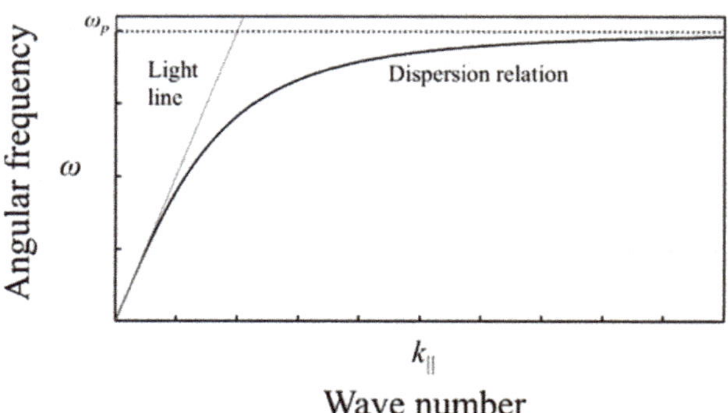

Figure 3. The dispersion relation of spoof surface plasmon polaritons (SSPP).

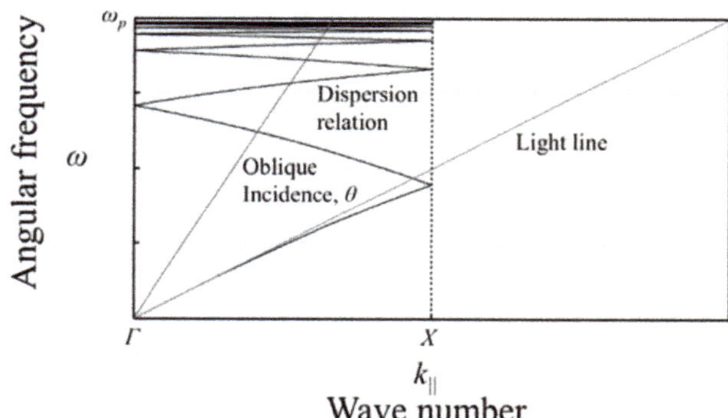

Figure 4. The real dispersion relation of SSPP with lattice scattering effects.

3. Electromagnetic Numerical Analyses of The MHA and Electromagnetic Distortion

In Section 2, we overviewed the physical basis of SSPP propagation and its dispersion, which showed that its frequencies are in fact caused and assured by the MHA. The concentration of its fields around the propagation interface was also pointed out, leading to potential applications of this propagation mode to 2D permittivity imaging that requires field uniformity over the area to be surveyed. However, the SSPP itself cannot provide us with local information on its responses to an electric field or permittivity, so in this section we confirm the effects of defect introduction on SSPP propagation.

Electromagnetic numerical analyses were performed to reconfirm the electromagnetic properties of the MHA predicted in Section 2 and to specify the electromagnetic field distortion on the MHA by a needle-like conductor. Figure 5 shows the analytical model using an electromagnetic simulator (HFSS R18, Ansys, Canonsburg, PA, USA). In this simulation model, the MHA is made of copper and its thickness is 2 mm. The size of the embedded waveguide is 2 mm × 1.5 mm and the period in both the x- and y- directions is 3 mm. The coordinate notations used here correspond to those in Section 2. Periodic boundary conditions applied to the sides of the model yield a hypothetical infinite area of SSPP propagation, and the incident surface (underside, port 1) is on the side opposite the receiving surface (topside, port 2) along the long side of the model, which explains the macroscopic wave propagation; they are defined as a Floquet port. Assuming the dispersion relation shown in Figure 4 for the MHA, SSPP modes exhibit resonant frequencies in a range lower than the cutoff frequency. To confirm this prediction, S-parameters (S_{11} and S_{21}) are analyzed under the above conditions in our numerical calculation, as shown in Figures 6 and 7.

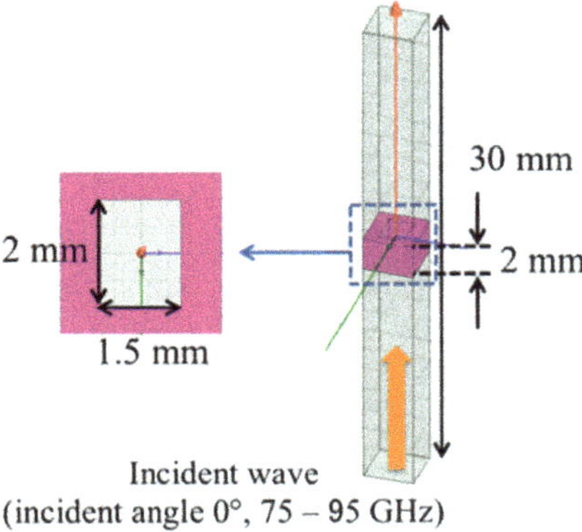

Figure 5. The MHA model for numerical analysis of the frequency characteristics of reflectance and transmittance.

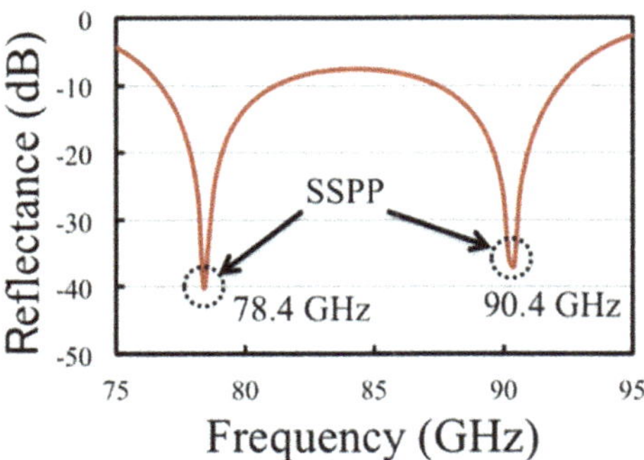

Figure 6. The frequency dependence of the reflectance of the MHA.

In Figures 6 and 7, the frequency spectrum of the transmission rate has peaks around 78.4 GHz and 90.4 GHz. This indicates the existence of at least two resonant frequencies in the MHA. These two resonances coincide with the intersection points of the dispersion relation of the SSPP and the light line in Figure 4; at these frequencies, energy conversion from the mode in free space propagation to the SSPP mode is smooth, thanks to good wave matching conditions. This phenomenon was termed extraordinary transmission as one of the features of an MHA [17–21]. In other words, at 78.4 GHz and 90.4 GHz, SSPP generation/propagation is possible where SSPP modes are able to uniformly concentrate electric fields on and in the vicinity of MHAs due to the multiple lattice scattering described in Equation (18).

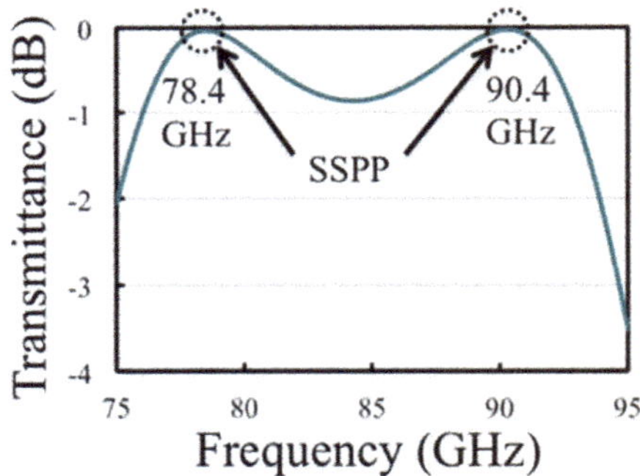

Figure 7. The frequency dependence of the transmittance of the MHA.

In this uniform 2D field, localized distortion is induced by inserting a needle-like conductor into a hole of the MHA. Electric fields are distorted strongly at sharp conductive boundaries, a phenomenon known as the near field effect [30], unlike wave propagation that includes a dispersion relation. Such distorted electric fields decay spatially with a wavelength-order attenuation constant. For this reason, another set of two analytical models (shown in Figure 8) was tested in our simulation runs. One has the MHA equipped with a conductive probe, and the other has only the MHA with its periodic 2D structure. That is, although the shape of both MHAs is the same as that in the model in Figure 5, the model has 5×5 holes in order to observe the difference in electric field distributions between case 1, insertion of a conductive probe in the MHA, and case 2, removal of the conductive probe. Note that the model of a conductive probe is positioned in or near a center hole. The probe has a cone shape, with a radius at the bottom plane of 0.5 mm and a height of 6 mm. Under these conditions, by using the model as shown in Figure 8, the electric field distributions of the two cases were monitored at 78.4 GHz, which is a candidate for the SSPP frequency near the resonant frequency observed in Figure 6. The results of the distributions in the two cases and their numerical comparison at z = +0.5 mm and −0.5 mm, are shown in Figure 9.

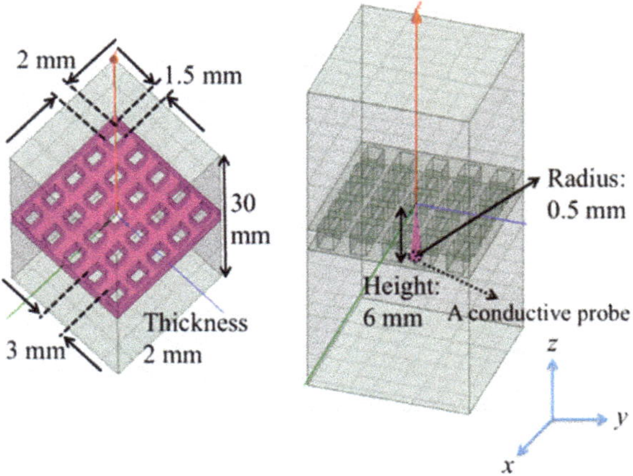

Figure 8. The two models for numerical analyses of the electric field distribution around the MHA.

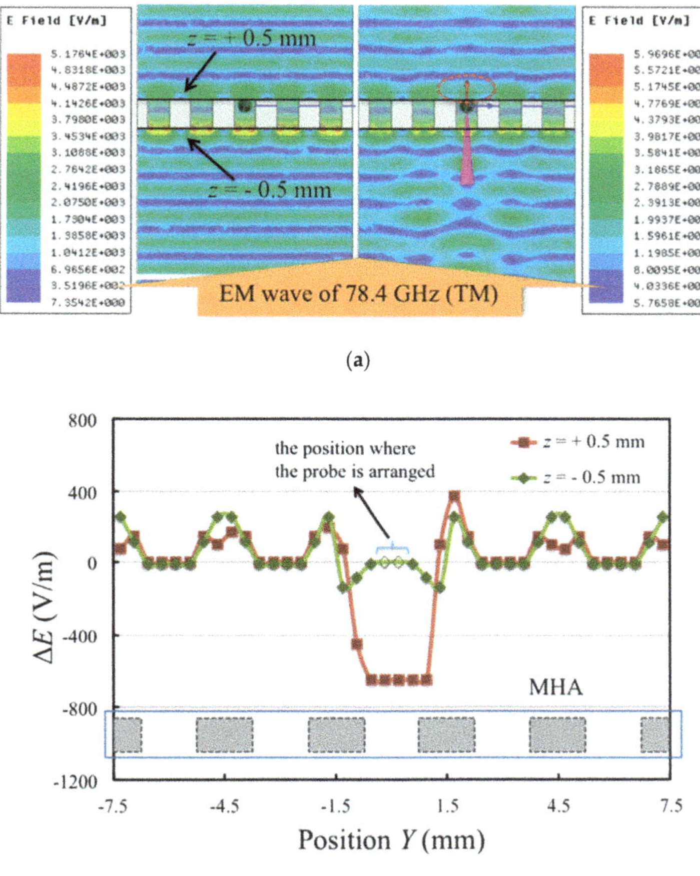

(**a**)

(**b**)

Figure 9. (**a**) The two electric field distributions around the MHA without (left) and with (right) a conductive probe; (**b**) the numerical results of the two cases ($z = +0.5, -0.5$ mm), which is the difference between the distributions with and without a conductive probe.

From Figure 9a, in the case on the left, electric fields are concentrated around the MHA, which is unlike the electric fields of the waveguide mode because those fields are concentrated not around holes but in holes. However, the results indicate that the fields are like compression waves, strongly affected by existing holes. Since the distribution of the fields around the MHA is symmetrical in the y-z plane, the electric-field distributions are partial components of an SSPP. This fact is consistent with a feature of SSPPs in which the mode expands its field on the side opposite the incident side when the mode is excited at an incident side. This result shows that the fields distribute by a coupling phenomenon between the SSPP mode and transmission waves. On the other hand, in the case on the right, looking at the electric distribution in the red circle, it was found that the electric field distributions are less concentrated than the distributions at the same region in the case on the left, while field distributions of the other holes in the case on the right are at the same level as those in the case on the left. These facts indicate that the electric fields in the region with the red circle are locally distorted by a conductive probe. Consequently, the results specifically show that localized electromagnetic distortions can be generated by inserting a conductive probe in a hole of an MHA.

From Figure 9b, as shown in the numerical results in the case of $z = +0.5$ mm in Figure 9a, it was found that there are significant differences in the region around a conductive probe. This is because the electric field distribution on the MHA is deformed by the insertion of a conductive probe, as displayed in Figure 9a. On the other hand, from the numerical results in the case of $z = -0.5$ mm, it was found that the difference between the two electric field distributions under the MHA is much smaller than the one on the MHA. This might be explained by the suggestion that the field distributions around the MHA are only distorted by a near-field effect. From these results, the differences in the two cases are potentially sufficient to detect the information in the deformed distributions because in practice, these differences are detected as signal intensities related to the square of the electric field.

The discussion above suggests that by taking the difference between the electric-field distributions of the holes when a conductive probe is inserted, and when not, local signals in the limited vicinity of each hole of the MHA can be obtained with detectable intensities and enhanced spatial resolution.

4. Two-Dimensional Imaging Experiments Using MHA and a Conductive Probe

In Section 3, the numerical results showed that an SSPP propagates in the MHA when electromagnetic waves at optimal frequencies enter the plate, and that its propagation can be locally deformed by a conductive metallic probe. To detect wave media information at the deformation point where electromagnetic waves enter the MHA, scattered waves generated apart from reflected waves play important roles in the frequency range, which includes the incidence of optimal waves. In the experiments discussed in this section, we used scattered waves, which exhibit the characteristics of local electromagnetic distribution [31]. Transmitter and receiver antennae, and a conductive probe are the main diagnostic tools manipulated for signal detection.

We performed experiments on one- and two-dimensional imaging using an MHA with structural parameters that were the same as those in the analytical model, and the MHA was equipped with a scanning conductive probe. After the one-dimensional imaging experiments, two-dimensional imaging experiments were performed using conductive and dielectric samples. Finally, we demonstrated a biomedical diagnosis in the case of a rat lung by using the system. The experimental system is shown in Figures 10–12, with a schematic view of the corresponding MHA and the conductive probe.

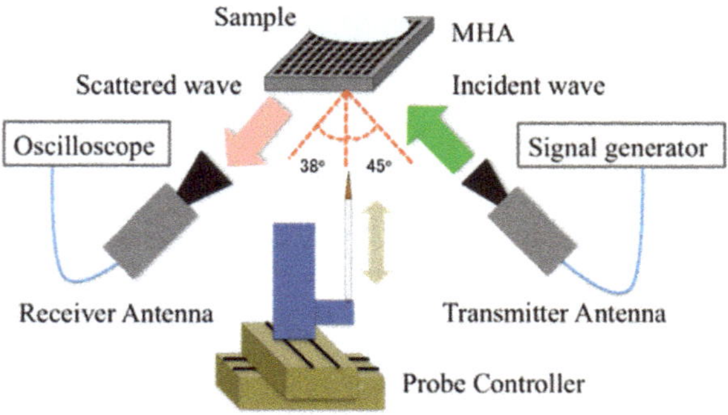

Figure 10. The experimental system for one- and two- dimensional imaging experiments.

As suggested by the above analyses, we searched for the most appropriate frequency for electromagnetic-wave experiments in the optimal bands for SSPP propagation, by using one-dimensional imaging experiments, which gave us simple and straightforward evidence as the conductive probe scanned along one line. With the incidence of electromagnetic waves at 78–86 GHz from a transmitter antenna into the structure shown in Figure 11, SSPPs form on the MHA and scattered waves are emitted, partly at the reflection angle of 38 degree. For a specific diagnostic sequence, two signals detected by a receiver antenna were recorded; in case 1: the conductive probe is inserted into a hole of the MHA, and in case 2: the conductive probe is removed from the MHA. By repeating the process, the signals in case 1 and case 2 of all the used holes shown in Figure 13 were obtained. Subtracting the signal of case 2 from that of case 1 for each hole, the differential signals of all the holes create one signal data set with a spatial profile. The processes described above were conducted for the following two series; for data set A; a sample (cylindrical aluminum stick, radius 1.6 mm, height 110 mm) on the MHA and for data set B; the sample was removed from the MHA. Finally, the one-dimensional imaging map was completed by subtracting the signals of data set A from those of data set B. The sample, and the results as an example are shown in Figure 13.

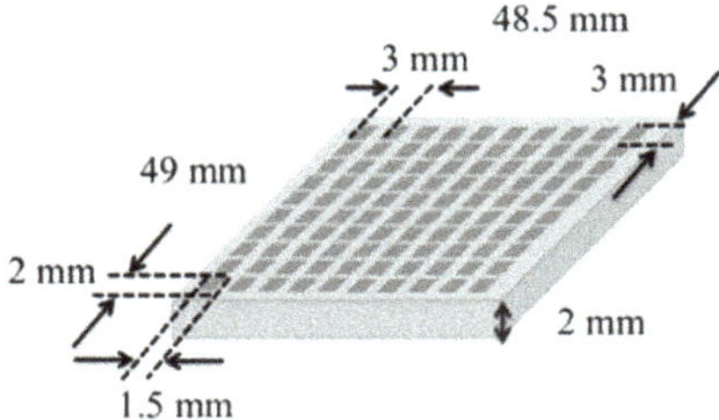

Figure 11. MHA of the imaging experiments.

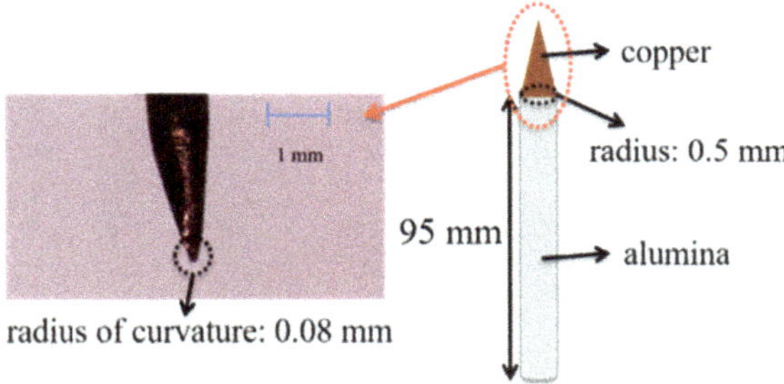

Figure 12. The conductive probe for localized electromagnetic distortion.

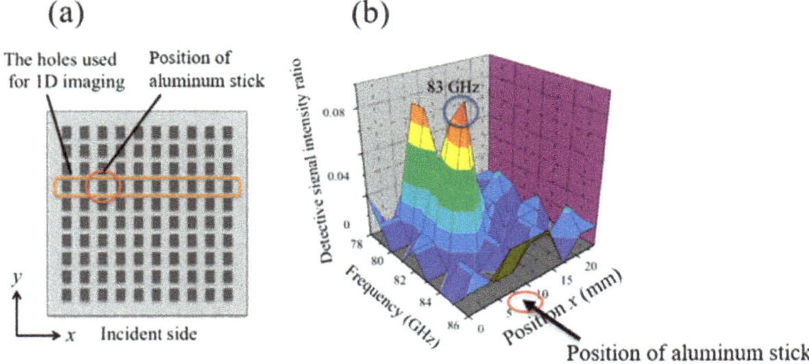

Figure 13. (**a**) The holes used for one-dimensional imaging; (**b**) the result of one-dimensional imaging for incident waves of 78–86 GHz.

Figure 13 shows that the detected signals of a conductive sample obtained at $x \sim 8$ mm, which corresponds to the location of the sample, are remarkable when electromagnetic waves in the frequency range of 78–80 GHz and around 83 GHz are incident. It also indicated that other waves cannot detect significant signals. Since the incident angle at each hole of the MHA differs slightly from the reflective angle in the system, detection capacity might depend on the frequency of incident waves to some degree. For this reason, using a longer distance from the MHA to the transmitter and receiver antennas may provide greater uniformity in SSPPs generated on the MHA. From the results shown in Figure 13, we fixed the operation frequency at 83 GHz.

After the selection of the wave frequency based on the one-dimensional experiment, we performed experiments for two-dimensional permittivity imaging. Three samples with different permittivity values were provided, and we used them for two-dimensional imaging experiments, using an experimental system similar to that shown in Figures 10–12. Similar to the one-dimensional imaging experiments, the samples were set up on the MHA, and optimal signals were assured by executing the process described above (two procedures for data sets A and B in which cases 1 and 2 are for profiling). The samples included one conductive material, copper sticks, and a dielectric solid and liquid, an alumina stick and DMSO (dimethyl sulfoxide). The values of relative permittivity of alumina and DMSO are 9.9 and 43 − j10.9, respectively, in the millimeter-wave band. DMSO works as a biomaterial phantom in our experiment, since its permittivity is nearly equal to that of the human body [32]. To contain the liquid DMSO in the setup, an acrylic capsule array was used as a supporting experimental tool. Therefore, when comparing signals, in data set A the sample was on and in; in data set B, the sample was removed from the MHA and an acrylic capsule array was set up in the same position on the MHA in both cases. The results of the two-dimensional imaging experiments for these three types of sample are shown in Figures 14–17.

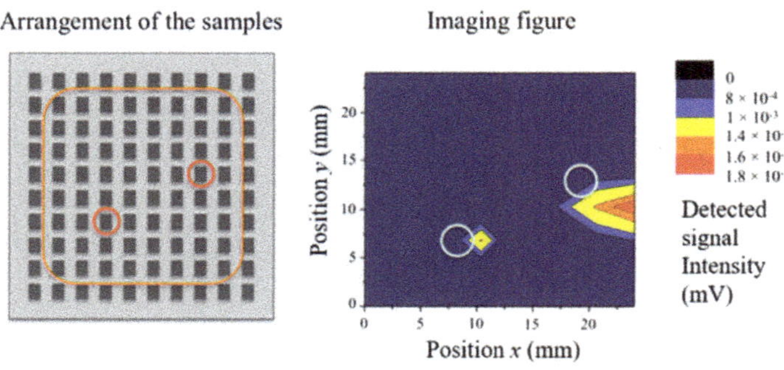

Figure 14. Two-dimensional imaging of conductive samples (two copper sticks).

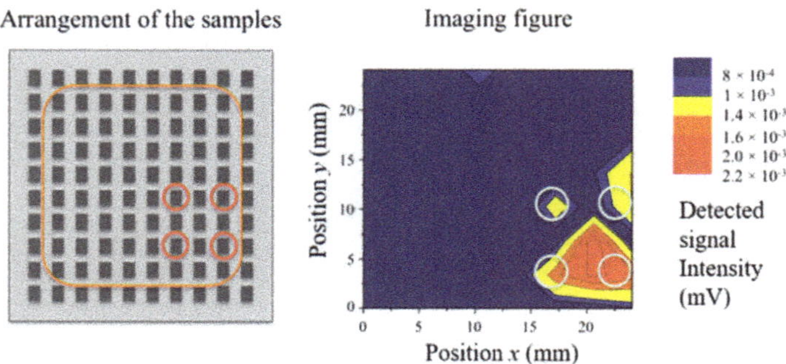

Figure 15. Two-dimensional imaging of conductive samples (four copper sticks).

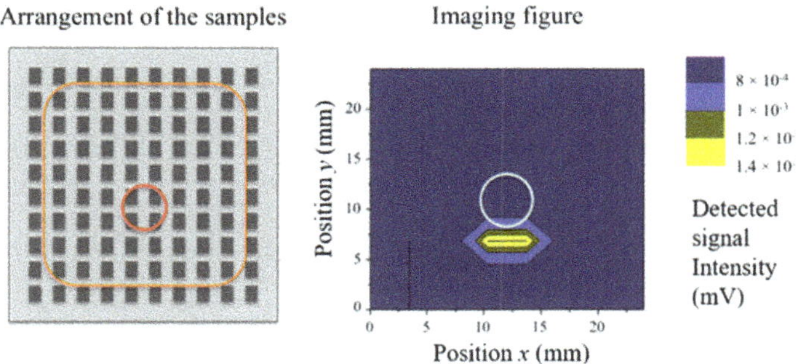

Figure 16. The two-dimensional imaging of dielectric samples (an alumina stick).

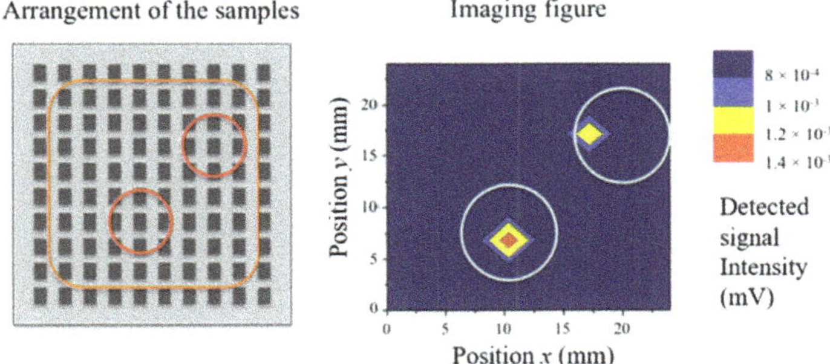

Figure 17. Two-dimensional imaging of dielectric samples (DMSO).

In Figures 14 and 15, although the positions of the detected signals are slightly different from the actual positions of the samples, the existence of the conductive samples was clearly detectable and related to their position. Also, in Figures 16 and 17, the results confirm the detection of dielectric samples at almost their proper positions even when the positions of the samples varied, though DMSO is a lossy material. These results show the effectiveness of the use of concentrated electric distributions for permittivity signal detection in this scheme, which differs from conventional methods of signal detection in the microwave band. Also, comparing the results of the conductive samples (Figures 14 and 15) to those of the dielectric samples (Figures 16 and 17), the positions of the detected signals of the former are less accurate than those of the latter. Considering the fact that the electric distributions of SSPPs are distorted to a greater extent by conductors than by dielectrics, the detection accuracy in the latter case may be superior to that in the former case. Furthermore, Figures 14 and 15 suggest that the detection accuracy for samples near corners is low in comparison with that of samples on inner locations. Since, in these experiments, the MHA has a limited size and SSPPs on the MHA are reflected at its corners, the electric distribution of SSPPs near corners tends to be non-uniform. Therefore, the distributions in the vicinity of its corners are not stable, and the detection accuracy of samples on corners tends to be low. By using an MHA with a larger number of holes, the detection capacity might be sufficiently accurate for practical use. On the other hand, the maximum values of the detected signal intensities of conductive samples are more intense than those of dielectric samples. In general, electric fields are affected more by conductors than dielectrics because of the boundary condition on the surface of conductors. Since DMSO has the large value of the imaginary part, it is consistent with the fact that the average value of the detected signal intensities in the case of the alumina sample was larger than that of DMSO. However, sufficient signal intensities were monitored in both cases for practical imaging, since the intensities satisfied the dynamic ranges of dB-order in the experiments. The results in all the cases confirm that the proper signals are obtainable at a mm-order distance by using the proposed method. This is an advantage over the conventional sensing method in which lasers or other strong light sources are required in the optical range.

The above discussions support the feasibility of a biomedical diagnosis that uses the proposed system. Finally, the result of the biomedical diagnosis in the case of a rat lung is reported. As shown in Figure 18, the results confirm a detection of the sample in more than half of the area. Although there is some slight signal misdetection, it shows that two-dimensional dialectic responses of the biomedical sample can be obtained in the proposed system.

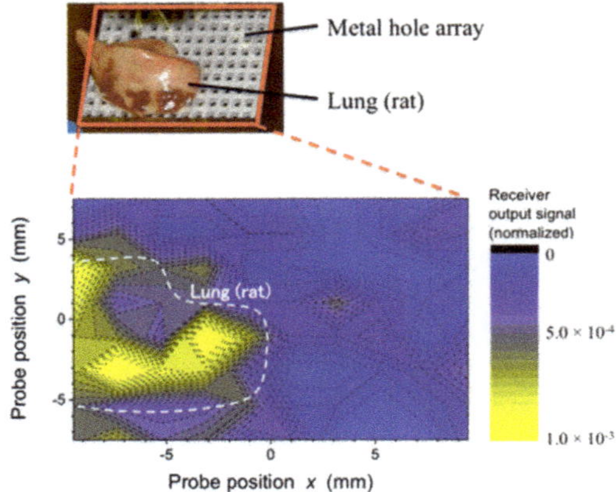

Figure 18. The result of biomedical diagnosis by using an MHA with rectangular holes (the size is 1.8 mm × 1.4 mm, the periods of each direction are 2.5 mm, 2 mm and the thickness is 2.3 mm) in the case of a rat lung.

The above discussions and results indicate that the proposed method can be used for 2D imaging of permittivity distribution, and may be useful for biomedical diagnosis after optimization as a medical facility tool.

5. Conclusions

A 2D imaging method is proposed in this study, based on a meta-structure with a scanning defect, using a metal hole array (MHA) and a conductive probe (a needle-like conductor) in the millimeter-wave range. In order to validate the proposed method, a theoretical model of SSPP generation on an MHA was introduced, numerical electromagnetic analyses of localized distortions of the electric fields on and around the MHA were conducted, and one- and two-dimensional imaging experiments using conductive and dialectic samples verified the theoretical predictions.

In Section 2, the dispersion relation of SSPPs in the case where the MHA has rectangular holes with oblique incidence of sampling waves was derived, which indicated the physical phenomenon of SSPP generation on an MHA. In Section 3, with the use of HFSS, the transmission and reflection properties of an MHA were analyzed when millimeter waves were injected into the structure, and this analysis implied that SSPP modes exist under the cut-off frequency in our MHA model. In addition, the electromagnetic distributions around an MHA with an inserted conductive probe were compared with those of an MHA without the probe, confirming the difference in the electric distributions around the MHA in the two cases. Specifically, insertion of a conductive probe into an MHA can be useful for localized electromagnetic distortion. In Section 4, using an MHA and a conductive probe, a one-dimensional experiment was performed and confirmed the frequency dependency of the detected signal intensity in the range of 78–86 GHz. This showed that a wave at 83 GHz can be monitored, producing the most intense level of detected signals. Then, two-dimensional imaging experiments with conductive samples (copper sticks) and dielectric samples (an alumina stick and DMSO (liquid)) were performed with the use of optimal waves at 83 GHz. Consequently, it was confirmed that the proposed measurement system allows the detection of the positions of conductive and dielectric samples by comparing the intensity levels of reflected signals with and without the samples. Finally, we demonstrated a biomedical diagnosis in the case of a rat lung by using the system. The result shows that although there is slight signal misdetection, two-dimensional dialectic responses of the biomedical sample can be obtained using the proposed method. Therefore, the proposed method has the potential for use in the two-dimensional imaging of permittivity distribution, such as in the biomedical task of localized tumor detection.

Author Contributions: G.I. performed the experiments and theoretical analyses shown here and wrote the paper by drawing the figures. O.S. conceived of and designed the research activities, as well as wrote the paper. Y.H. accelerated the study by presenting the technical issues in the medical point of view.

Funding: This research was partly funded by Project for Kyoto Bio-industry Creation & Support, and by JSPS Kakenhi with grant number 18H03690.

Acknowledgments: The author thanks Y. Nishio and A. Iwai at Kyoto University for their fruitful comments on this study.

Conflicts of Interest: The authors declare no conflict of interest.

References

1. Hasch, J.; Topak, E.; Schnabel, R.; Zwick, T.; Weigel, R.; Waldschmit, C. Millimeter-wave Technology for Automotive Radar sensors in the 77 GHz Frequency Band. *IEEE Trans. Microw. Theory Tech.* **2012**, *60*, 845–860. [CrossRef]
2. Theofanopoulos, P.C.; Trichopoulos, G.C. Modeling of mm W and THz Imaging Systems Using Conjugate Field Coupling. *IEEE Antennas Wirel. Propag. Lett.* **2018**, *17*, 213–216. [CrossRef]
3. Ghasr, M.T.; Horst, M.J.; Dvorsky, M.R.; Zoughi, R. Wideband Microwave Camera for real-Time 3-D Imaging. *IEEE Trans. Antennas Propag.* **2017**, *65*, 258–268. [CrossRef]
4. Long, F.; He, M.; Shi, H.C.; Zhu, A.N. Development of evanescent wave all-fiber immunosensor for environmental water analysis. *Biosens. Bioelectron.* **2008**, *23*, 952–958. [CrossRef] [PubMed]
5. Katyal, J.; Soni, R.K. Localized Surface Plasmon Resonance and Refractive Index Sensitivity of Metal-Dielectric-Metal Multilayered Nanostructure. *Plasmonics* **2014**, *9*, 1171–1181. [CrossRef]
6. Tiwari, K.; Sharma, S. Surface plasmon based sensor with order-of-magnitude higher sensitivity to electric field induced changes in dielectric environment at metal/nematic liquid-crystal interface. *Sens. Actuators A* **2014**, *216*, 128–135. [CrossRef]
7. Zhu, Q.; Tan, W.; Wang, Z. The combined effect of side-coupled gain cavity and lossy cavity on the plasmonic response of metal-dielectric-metal surface plasmon polariton waveguide. *J. Phys. Condens. Matter* **2014**, *26*, 255301. [CrossRef] [PubMed]
8. Strachko, A.A.; Agranovich, V.M. To the theory of surface plasmon on metals covered with resonant thin films. *Opt. Commun.* **2014**, *332*, 201–205. [CrossRef]
9. Anker, J.N.; Hall, W.P.; Lyandres, O.; Shah, N.C.; Zhao, J.; Duyne, R.P.V. Biosensing with plasmonic nanosensors. *Nat. Mater.* **2008**, *7*, 442–453. [CrossRef]
10. Kreno, L.E.; Leong, K.; Farha, O.K.; Allendorf, M.; Duyne, R.P.V.; Hupp, J.T. Metal-Organic Framework Material as Chemical Sensors. *Chem. Rev.* **2012**, *112*, 1105–1125. [CrossRef]
11. Camden, J.P.; Dieringer, J.A.; Zhao, J.; Duyne, R.P.V. Controlled Plasmonic Nanostructures for Surface-Enhanced Spectroscopy and Sensing. *Acc. Chem. Res.* **2008**, *41*, 1653–1661. [CrossRef] [PubMed]
12. Pendry, J.B.; Moreno, L.M.; Vidal, F.J.G. Mimicking Surface Plasmon with Structured Surfaces. *Science* **2004**, *305*, 847–848. [CrossRef] [PubMed]
13. Woolf, D.; Kats, M.A.; Capasso, F. Spoof surface plasmon waveguide forces. *Opt. Lett.* **2014**, *39*, 517–520. [CrossRef] [PubMed]
14. Bhattacharya, S.; Shah, K. Multimodal propagation of the electromagnetic wave on a structured perfect electric conductor (PEC) surface. *Opt. Commun.* **2014**, *328*, 102–108. [CrossRef]

15. Shen, L. Effect of absorption on terahertz surface plasmon polaritons propagating along periodically corrugated metal wires. *Phys. Rev. B* **2008**, *77*, 075408. [CrossRef]

16. Ng, B.; Hanham, S.M.; Wu, J.; Dominguez, A.I.F.; Klein, N.; Liew, Y.F.; Breese, M.B.; Hong, M.; Maier, S.A. Broadband Terahertz Sensing on Spoof Plasmon Surfaces. *ACS Photonics* **2014**, *1*, 1059–1067. [CrossRef]

17. Miyamaru, F.; Hangyo, M. Anomalous terahertz through double-layer metal hole arrays by coupling of surface plasmon polaritons. *Phys. Rev. B* **2005**, *71*, 165408. [CrossRef]

18. Miyamaru, F.; Hayashi, S.; Otani, C.; Kawase, K. Terahertz surface-wave resonant sensor with a metal hole array. *Opt. Lett.* **2006**, *31*, 1118–1120. [CrossRef] [PubMed]

19. Miyamaru, F.; Kamijyo, M.; Hanaoka, N.; Takeda, M.W. Controlling extraordinary transmission characteristics of metal hole arrays with spoof surface plasmons. *Appl. Phys. Lett.* **2012**, *100*, 081112. [CrossRef]

20. Miyamaru, F.; Hangyo, M. Finite size effect of transmission property for metal hole arrays in subterahertz region. *Appl. Phys. Lett.* **2004**, *84*, 2742–2744. [CrossRef]

21. Miyamaru, F. Effect of hole diameter on terahertz surface-wave excitation in metal hole arrays. *Phys. Rev. B* **2006**, *74*, 153416. [CrossRef]

22. Brissova, M.; Fowler, M.J.; Nicholson, W.E.; Chu, A.; Hirshberg, B.; Harlan, D.M.; Powers, A.C. Assessment of Human Pancreatic Islet Architecture and Comparison by Laser Scanning Confocal Microscopy. *J. Histochem. Cytochem.* **2005**, *53*, 1087–1097. [CrossRef] [PubMed]

23. Klein, A.E.; Janunts, N.; Steinert, M.; Tunnermann, A.; Pertsch, T. Polarization-Resolved Near-Field Mapping of Plasmonic Aperture Emission by a Dual-SNOM System. *Nano Lett.* **2014**, *14*, 5010–5015. [CrossRef] [PubMed]

24. Itami, G.; Sakai, O. Symmetrical Estimation Method for Skin Depth-Control of Spoof Surface Plasmon by Using Dispersed Waves from Metallic Hole Array. *J. Appl. Phys.* under review.

25. Meltzer, C.C.; Price, J.C.; Mathis, C.A.; Greer, P.J.; Cantwell, M.N.; Houck, P.R.; Mulsant, B.H.; Ellezer, D.B.; Loprestl, B.; Dekosky, S.T.; et al. PET Imaging of Serotonin Type 2A Receptors in Late-Life Neuropsychiatric Disorders. *Am. J. Psychiatry* **1999**, *156*, 1871–1878. [PubMed]

26. Haim, S.B.; Murthy, V.L.; Breault, C.; Allie, R.; Sitek, A.; Roth, N.; Fantony, J.; Moore, S.C.; Park, M.A.; Kijewski, M.; et al. Quantification of Myocardial Perfusion Reserve Using Dynamic SPECT Imaging in Humans: A Feasibility Study. *J. Nucl. Med.* **2013**, *54*, 873–879. [CrossRef] [PubMed]

27. Juette, M.F.; Gould, T.J.; Lessard, M.D.; Mlodzianoski, M.J.; Nagpure, B.S.; Bennett, B.T.; Hess, S.T.; Bewersdorf, J. Three-dimensional sub-100 nm resolution fluorescence microscopy of thick samples. *Nat. Methods* **2008**, *5*, 527–529. [CrossRef] [PubMed]

28. Kubota, S.; Xiao, X.; Sasaki, N.; Kayaba, Y.; Kimoto, K.; Moriyama, W.; Kozaki, T.; Hanada, M.; Kikkawa, T. Confocal Imaging Using Ultra Wideband Antenna Array on Si Substrates for Breast Cancer Detection. *Jpn. J. Appl. Phys.* **2010**, *49*, 097001. [CrossRef]

29. Vidal, F.J.G.; Moreno, L.M.; Pendry, J.B. Surfaces with holes in them: New plasmonic metamaterials. *J. Opt. A Pure Appl. Opt.* **2005**, *7*, S97–S101. [CrossRef]

30. Sekimoto, K.; Takayama, M. Negative ion formation and evolution in atmospheric pressure corona discharges between point-to-plane electrodes with arbitrary needle angle. *Eur. Phys. J. D* **2010**, *60*, 589–599. [CrossRef]

31. Itami, G.; Akiyama, T.; Sakai, O.; Harada, Y. Localized electromagnetic distortion in 2D metal hole array and its application to imaging of permittivity distribution. In Proceedings of the Asia-Pacific Microwave Conference (APMC), Sendai, Japan, 4–7 November 2014.

32. Urata, M.; Kimura, S.; Wakino, K.; Kitazawa, T. Optimization of three layer tapered coaxial line applicator for cancer thermotherapy by impedance matching of transition layer sections. In Proceedings of the IEEE International Symposium on Antennas and Propagation (APSURSI), Spokane, WA, USA, 3–8 July 2011.

 *electronics*

Article

A Dual-Band Compact Metamaterial Absorber with Fractal Geometry

Francesca Venneri, Sandra Costanzo * and Antonio Borgia

DIMES - Università della Calabria, 87036 Rende, Italy
* Correspondence: costanzo@dimes.unical.it; Tel.: +39-0984-494652

Received: 9 June 2019; Accepted: 6 August 2019; Published: 8 August 2019

Abstract: A fractal absorber based on a metamaterial configuration is proposed for dual-frequency operation within the UHF band. The miniaturization skills of the proposed fractal shape are used to design a dual-band metamaterial absorber cell with reduced size ($<\lambda/2$ at the two operating frequencies) and a very thin substrate thickness ($\cong\lambda/100$). A metamaterial absorber panel is realized and experimentally validated. Good agreements between full-wave simulations and measurement results are demonstrated.

Keywords: dual-band; fractals; microwave absorbers; UHF-RFID

---

1. Introduction

Electromagnetic absorbers are usually adopted to reduce unwanted reflections of an impinging wave within a prescribed frequency band. Several types of electromagnetic absorbers have been developed in literature [1–3] for various applications throughout the electromagnetic spectrum. Among the above designs, metamaterial absorbers (MAs) have been proposed in Reference [2] as giving small-volume and low-cost absorbing structures. They are composed of a set of periodic resonant metallic patches on a grounded dielectric layer. MAs can be properly synthesized to perform ideal absorption in the neighborhood of a given frequency. Since their first presentation [2], MAs have been employed in various applications [3], such as electromagnetic interferences reduction, multi-path mitigation of indoor RF-applications, and electromagnetic compatibility enhancement of electronics devices. Recently, MAs have been also proposed to operate at low microwave frequencies, usually requiring much more cumbersome absorption structures. In particular, some MA configurations have been designed in References [4–8] to improve the robustness of RFID systems working in the Ultra High Frequency (UHF) range.

In this paper, an ultra-thin fractal MA cell, first illustrated by the authors in Reference [9], is presented for dual-band operation within the UHF range. Even if other dual-band/multi-band MA configurations can be found in literature [10–17], this is the first example working at UHF frequencies. Unlike other multi-band MAs [10–17], the proposed fractal MA-cell allows to achieve closer absorbing peaks. Furthermore, the adopted single-layer structure offers reduced thickness ($\cong\lambda/107$) and very small unit cells ($\cong0.4\lambda$), with λ being the free-space wavelength at the smaller resonant frequency. In order to provide a preliminary experimental validation of the proposed absorber configuration, a MA panel operating within the UHF-band is tested. Very low reflection coefficient values (≤-22 dB) are detected, under normal incidence, at both operating frequencies, thus providing very high absorption rates ($\geq99\%$) in dual-band operation.

2. Dual-Frequency Fractal Unit Cell

The dual-band MA structure proposed in this paper is illustrated in Figure 1, where the black fill part is a conductor (35 µm-thick copper). It consists of fractal cells periodically distributed and printed on a thin grounded dielectric layer (Figure 1). The MA-unit cell is composed of two alternately arranged couple of Minkowski patches (Figure 1), each one synthesized to work in the neighborhood of a specific resonant frequency. The proposed MA structure is backed by a metallic sheet (i.e., the ground plane in Figure 1). For this reason, there is no transmission through the absorber panel, so we only need the MA reflection response (Γ) to correctly compute the absorption (i.e., A = 1 − $|\Gamma|^2$) of the proposed structure.

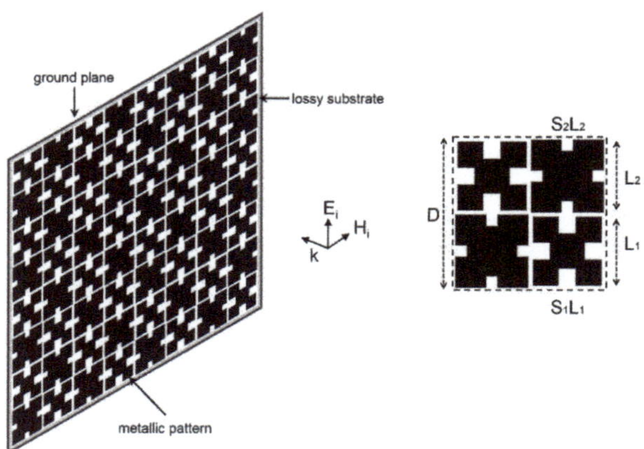

Figure 1. Fractal metamaterial absorber: dual-band unit cell.

The Minkowski fractal geometry, yet proposed by the authors in Reference [7] for designing a single-frequency UHF MA-cell, is composed by a starting L × L square element, which is properly changed by eliminating a smaller SL × SL square from the center of its sides. S is the scaling factor, varying from 0 up to 1/3.

The above fractal configuration allows to host an electrically longer resonator into a smaller unit cell [18,19], thus offering very appealing miniaturization skills. As a matter of the fact, the combined use of a smaller patch length L and a greater S-value allows to move down the patch resonance frequency, f_0. This last parameter, in fact, is inversely proportional to the effective side length L^{eff} of the patch (i.e., $f_0 \sim 1/L^{eff}$), which is approximately equal to L^{eff} = (1 + 2S) L [8]. So, the longer the effective fractal side is, the smaller the resonant frequency will be. Furthermore, in order to realize the perfect absorption condition at the desired resonant frequency f_0, both degrees of freedom of the adopted Minkowski shape, i.e., the length L and the inset size SL, can be exploited. In other words, the fractal sizes must be chosen to match the MA unit cell impedance, Z_{cell}, with the free-space impedance ξ_0 = 377 Ω, in correspondence of the design frequency f_0 (i.e., $|\Gamma(f_0)| = \left| \frac{Z_{cell}(f_0) - \xi_0}{Z_{cell}(f_0) + \xi_0} \right| \cong 0$, with Γ the reflection coefficient of the unit cell, see Figure 2a).

As demonstrated in the following, the proposed two-degree-of-freedom design technique simplifies the synthesis of a perfect MA-cell, giving simultaneous absorption and resonance conditions at the desired frequency, by tuning both fractal parameters L and S. Furthermore, the proposed geometry can be easily scaled to operate at different frequencies.

A test case demonstrating the above considerations is illustrated in Figure 2a. It represents the simulated reflection coefficient of a set of 868 MHz fractal MA cells, with progressively reduced sizes D, ranging from 0.3λ down to 0.15λ, with λ being the free-space wavelength. A commercial full-wave code is adopted (i.e., Ansoft Designer) by imposing the infinite periodic array boundary conditions.

Results reported in Figure 2 show that a proper tuning of parameters L and S is able to guarantee an ideal absorption condition in all considered cases (i.e., $|\Gamma(f_0)| \cong 0$, namely $A(f_0) > 99\%$ at f_0 = 868 MHz, with $A(f) = 1 − |\Gamma(f)|^2$). In fact, even if the use of cell size D of reduced dimension forces smaller patch lengths L (Figure 2a), the use of greater values for the scaling parameter S allows to have a patch with an increased electrical length, thus giving a downshift of the resonance condition with respect to the desired working frequency $f_0 = 868$MHz (Figure 2a). From a physical point of view, the above behavior can be justified by the fact that the current path on the fractal patch is bent in correspondence of the inset SL (see Figure 3), thus the actual resonance length of the patch results to be lengthened by the SL notch.

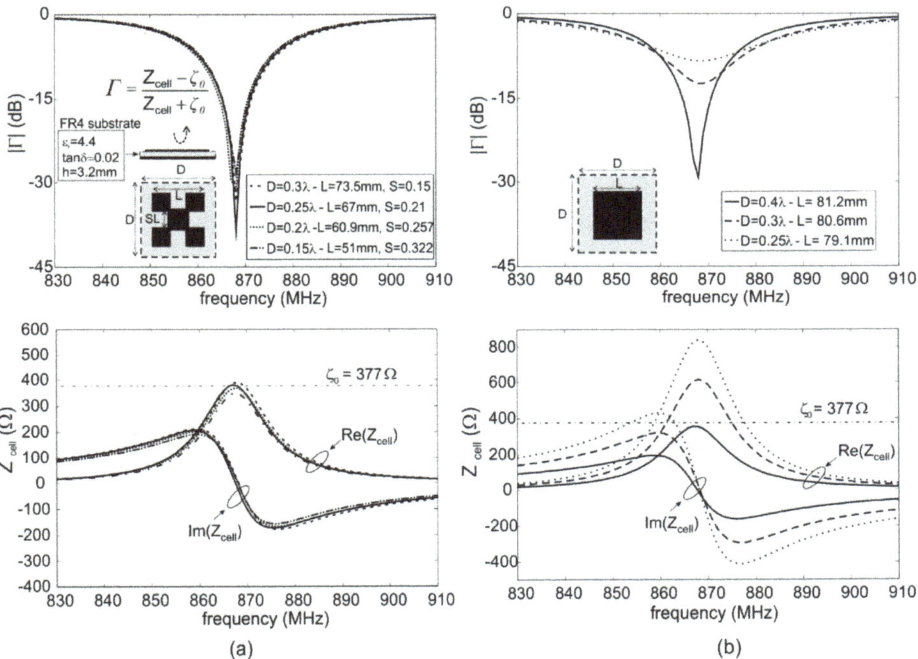

Figure 2. Reflection coefficient and unit cell input impedance vs. frequency for different cell sizes D: (**a**) fractal-based cell; (**b**) square-based cell.

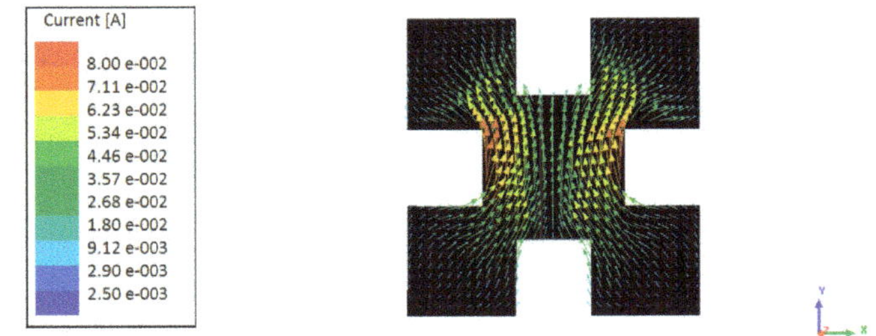

Figure 3. Current on the patch surface @ 868 MHz (D = 0.2λ – L = 60.9 mm, S = 0.257).

For the above reasons, the adopted fractal geometry allows to design more compact unit cells: for the examples depicted in Figure 2, a size reduction of about 63% is achieved with respect to the conventional square-shaped MA cell, printed on the same substrate (see Figure 2b).

Moreover, Figure 2b illustrates how the reflection coefficient $\Gamma(f_0)$ of the unit cell relative to the square-shaped MAs increases when a smaller unit cell size D is considered, thus leading to a nonoptimal absorption. This behavior is due to the fact that the synthesis of a square-based MA-cell is based on the use of only one degree of freedom, i.e., the patch length L that is commonly chosen to fix the unit-cell resonance at the design frequency f_0. No degree of control over the unit cell impedance is available, which instead depends on the following preset parameters: the unit cell size D, the resonant patch dimension L, and the substrate features (i.e., thickness, dielectric constant, loss tangent). As a matter of the fact, in the case of the square-based cells depicted in Figure 2b, the use of a smaller array grid spacing leads to a larger resonant input impedance value (i.e., Re(Z_{cell})), mainly due to the higher capacitive coupling between adjacent patches [8,20] that causes a progressively mismatching and a consequent reduction of MA absorption abilities (Figure 2b).

The above miniaturization capabilities are exploited in this paper to realize a dual-band feature by simply including two couples of fractal resonators in the same cell. The design process of the proposed dual-band cell follows the design rules described in Reference [8] for the case of a single resonant MA. Each fractal patch is synthesized to realize and ideal absorption, i.e., to have a matching condition between the impedances relative to the MA cell and that of free space [8] at the design frequencies f_1 and f_2 (i.e., $|\Gamma(f_1)| \cong |\Gamma(f_2)| \cong 0$).

A MA unit cell, printed on a FR4 substrate (ε_r = 4.2, tanδ = 0.015, h = 3.2 mm), is designed by adopting the layout depicted in Figure 1. A commercially available full-wave code, based on the infinite array condition, is applied, with an assumed normal incident plane wave (θ_{inc} = 0°). As demonstrated in Figure 4 (computed reflection coefficient of the cell), two reflection nulls are achieved (i.e., $|\Gamma(f_1)| \cong -22$ dB, $|\Gamma(f_2)| \cong -33$ dB) at the following UHF-frequencies: f_1 = 878 MHz and f_2 = 956 MHz. The above nulls correspond to a pair of absorption peaks, $A(f_1)$ and $A(f_2)$, greater than 99%. The synthesized MA cell is characterized by the following dimensions: (L_1, S_1) = (64.2 mm, 0.245), (L_2, S_2) = (68 mm, 0.155), D $\cong$ 0.4λ, where λ is the free space wavelength at the lower frequency f_1. As a further analysis, good angular stability is demonstrated for both TE (Figure 4a) and TM (Figure 4b) polarizations up to θ_{inc} = 60°. To this end, the reflection coefficients vs. frequency are computed for different incidence angles θ_{inc}.

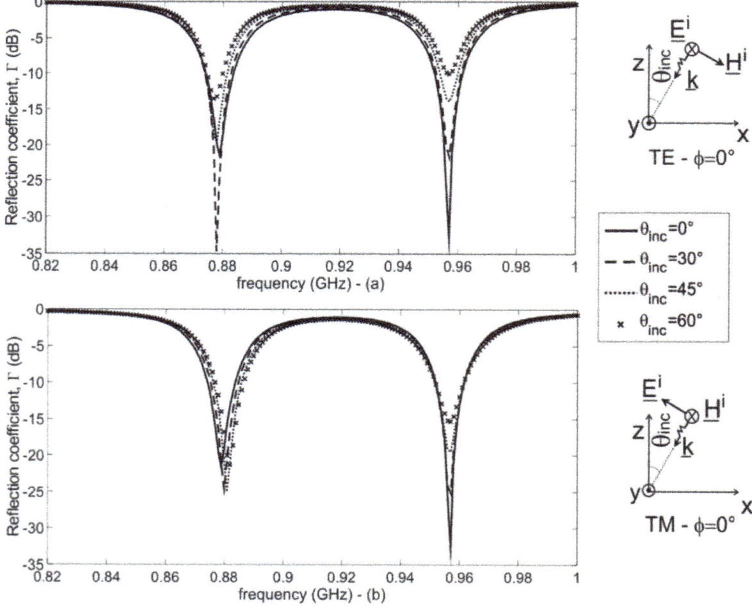

Figure 4. Reflection coefficient vs. frequency of the synthesized dual-band metamaterial absorber (MA) cell for different incidence angles (ε_r = 4.2, tanδ = 0.015 and h = 3.2 mm–D $\cong$ 0.4λ @ f_1): (**a**) TE-polarization (**b**) TM-polarization.

Figure 4 shows a little frequency shift in correspondence of the first resonance, in the case of a TM polarization. Moreover, a reflection amplitude ranging from −34 dB up to −14 dB, at f_1 (i.e., $A(f_1) \geq 96\%$), and from −32 dB up to −10.2 dB, at f_2 (i.e., $A(f_2) \geq 90.5\%$), is achieved for a TE polarized incident plane wave (Figure 4a). On the other hand, a reflection coefficient ranging from −25 dB up to −22 dB, at f_1 (i.e., $A(f_1) \geq 99.4\%$), and from −33 dB up to −15.3 dB, at f_2 (i.e., $A(f_2) \geq 97\%$), is obtained for the TM polarization (Figure 4b). As demonstrated in Reference [20], the angular stability of MA panels mainly depends on their thickness. In particular, the use of a very thin dielectric substrate, as in the case of the proposed MA configuration, is able to reduce the incidence angle effect [20].

In conclusion, the proposed fractal MA-cell offers reduced thickness ($\cong \lambda/100$) and small unit cells ($\cong 0.4\lambda$) at both operating frequencies, with respect to the most existing dual-band MA configurations (see Table 1). Moreover, it allows to achieve very close absorbing peaks by virtue of combining two distinct pairs of resonators both characterized by two degrees of freedom (i.e., the fractal sizes (L_i, S_i) $-i = 1, 2$, see Figure 1). In fact, the latter can be specifically tuned to obtain very close resonance lengths (i.e., $L_i^{\text{eff}} = (1 + 2S_i) L_i$, $i = 1, 2$, see Figure 1), namely very close resonant frequencies ($f_i \sim 1/L_i^{\text{eff}}$), assuring at the same time perfect absorption at both resonant frequencies. A quick demonstration of the above statements is reported in Figure 5, showing the design of a MA cell with closer absorption peaks. In particular, the 1st absorption frequency is fixed to the value $f_1 = 880$ MHz, while the 2nd absorption frequency is shifted down from 970 MHz to 925 MHz, by changing both parameters L_2 and S_2 in order to increase the effective length L_2^{eff}, thus guaranteeing at the same time a well matched impendence, namely a good absorptivity (i.e., $\Gamma(f_2) \leq -25$ dB $\Rightarrow A(f_2) \geq -99.6\%$).

Table 1. Comparison with previously proposed dual-band MAs.

Configuration	f_1 (GHz)	f_2 (GHz)	Cell Periodicity	Thickness
Present work	0.878	0.956	$\cong 0.4\lambda_1$ ($0.44\lambda_2$)	$\lambda_1/107$ ($\lambda_2/98$)
[10]	17	18	$\cong 0.5\lambda_1$ ($0.54\lambda_2$)	$\lambda_1/44$ ($\lambda_2/42$)
[11]	11.15	16	$\cong 0.45\lambda_1$ ($0.64\lambda_2$)	$\lambda_1/54$ ($\lambda_2/38$)
[12]	5.6	6	$\cong 0.38\lambda_1$ ($0.4\lambda_2$)	$\lambda_1/36$ ($\lambda_2/33$)
[13]	4.42	5.62	$\cong 0.44\lambda_1$ ($0.56\lambda_2$)	$\lambda_1/42$ ($\lambda_2/33$)

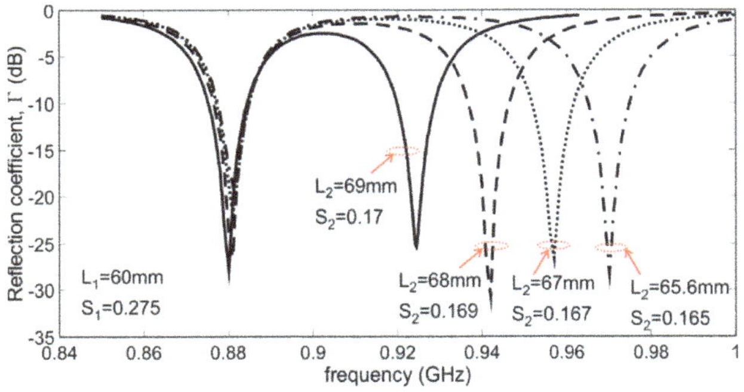

Figure 5. Reflection coefficient vs. frequency of a synthesized dual-band MA for different (L_2, S_2)-values ($\varepsilon_r = 4.2$, $\tan\delta = 0.015$ and $h = 3.2$ mm–D $\cong 0.4\lambda$ @f_1).

3. Experimental Validation

To perform a preliminary experimental validation of the proposed dual-band configuration, a 125 mm × 125 mm absorber panel, composed by 18 × 18 fractal elements, is realized and tested in the anechoic chamber of the Microwave Laboratory of the University of Calabria, which is equipped for both near-field and far-field measurements [21,22]. For the specific application, a far-field experimental setup (Figure 6), including a transmitting and a receiving broadband horn antennas, is adopted to measure the reflection coefficient of the absorbing panel. Both horn antennas, connected to a vector network analyzer (VNA), are placed at a distance X from the panel satisfying the far-field condition (i.e., $X > 2D^2/\lambda_0$, where D is the diagonal length of the horn antennas). The horns are rotated on their own axis in order to measure the panel reflection coefficient, for incident signals in both TE and TM polarization. The reflection coefficient of the MA panel (Figure 7) is measured under normal incidence (i.e., $\theta_{inc} \cong 0°$). An absorptivity greater than 96% and 99.4% is demonstrated at the frequencies f_1 and f_2, respectively, for both polarizations (Figure 7). Furthermore, a very good agreement is achieved between experimental and simulation data (Figure 7), thus demonstrating the effectiveness of the proposed dual-band fractal MA to realize perfect absorption. Finally, very low sensitivity to the angular variation of the incident wave is experimentally checked for both TE and TM polarizations (Figure 7).

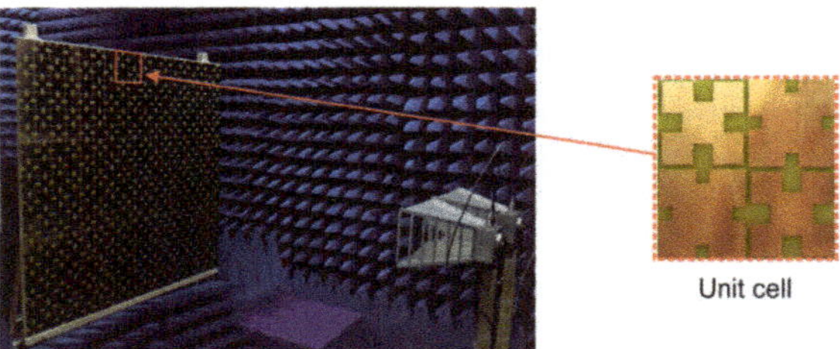

Figure 6. Free-space measurement setup with a particular of the fabricated MA absorber.

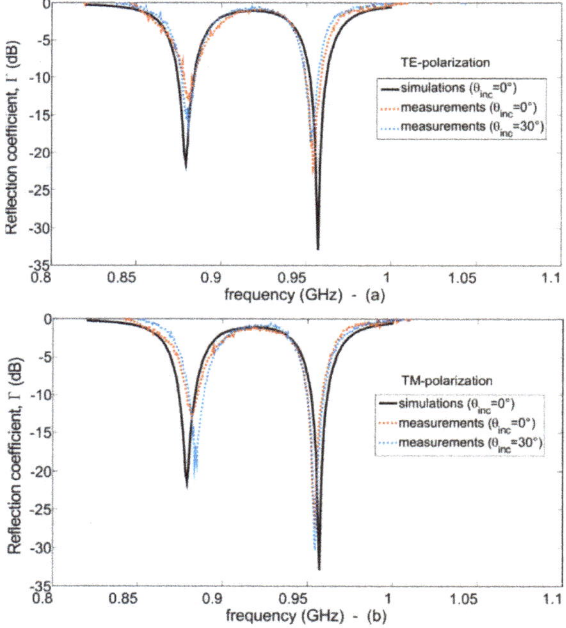

Figure 7. Comparison between measured and simulated reflection coefficients vs. frequency: (**a**) TE-polarization; (**b**) TM-polarization.

4. Conclusions

The compactness capabilities of fractal geometries have been considered in this paper to design a dual-band microwave absorber operating within the UHF band. Very high absorption peaks (>99%) have been simulated in the neighborhood of the desired working frequencies (i.e., 878 MHz and 956 MHz). A fractal MA panel has been prototyped and experimentally validated, confirming the capabilities of the proposed configuration in realizing the perfect absorption condition under the dual-frequency modality, thus saving thickness and size as compared to standard microwave absorbing configurations.

Author Contributions: The individual contributions of authors are specified as follows: Conceptualization, S.C. and F.V.; methodology, S.C. and F.V.; software, F.V.; validation, S.C., F.V., and A.B.; writing—original draft preparation, F.V. and A.B.; writing—review and editing, S.C.; supervision, S.C.

Funding: This research received no external funding.

Conflicts of Interest: The authors declare no conflict of interest.

References

1. Munk, B.A. *Frequency Selective Surfaces: Theory and Design*; Wiley Interscience: New York, NY, USA, 2000.

2. Landy, N.I.; Sajuyigbe, S.; Mock, J.J.; Smith, D.R.; Padilla, W.J. Perfect metamaterial absorber. *Phys. Rev. Lett.* **2008**, *100*, 207402. [CrossRef] [PubMed]

3. Watts, C.M.; Liu, X.; Padilla, W.J. Metamaterial electromagnetic wave absorbers. *Adv. Mater.* **2012**, *24*, 98–120. [CrossRef] [PubMed]

4. Okano, Y.; Ogino, S.; Ishikawa, K. Development of optically transparent ultrathin microwave absorber for ultrahigh-frequency RF identification system. *IEEE Trans. Microw. Theory Tech.* **2012**, *60*, 2456–2464. [CrossRef]

5. Costa, F.; Genovesi, S.; Monorchio, A.; Manara, G. Low-cost metamaterial absorbers for sub-GHz wireless systems. *IEEE Antennas Wirel. Propag. Lett.* **2014**, *13*, 27–30. [CrossRef]

6. Costa, F.; Genovesi, S.; Monorchio, A.; Manara, G. Perfect metamaterial absorbers in the ultra-high frequency range. In Proceedings of the International Symposium on Electromagnetic Theory, Hiroshima, Japan, 20–24 May 2013.

7. Zuo, W.; Yang, Y.; He, X.; Zhan, D.; Zhang, Q. A miniaturized metamaterial absorber for ultrahigh-frequency RFID system. *IEEE Antennas Wirel. Propag. Lett.* **2017**, *16*, 329–332. [CrossRef]

8. Venneri, F.; Costanzo, S.; Di Massa, G. Fractal-shaped metamaterial absorbers for multireflections mitigation in the UHF band. *IEEE Antennas Wirel. Propag. Lett.* **2018**, *17*, 255–258. [CrossRef]

9. Venneri, F.; Costanzo, S.; Di Massa, G. Multi-band designs of fractal microwave absorbers. In *Advances in Intelligent Systems and Computing, Proceedings of the WorldCIST'18, Naples, Italy, 27–29 March 2018*; Springer: Cham, Switzerland, 2018; Volume 746.

10. Singh, D.; Srivastava, V.M. Dual resonance shorted stub circular rings metamaterial absorber. *Int. J. Electron. Commun. (AEÜ)* **2018**, *83*, 58–66. [CrossRef]

11. Li, M.H.; Yang, H.L.; Hou, X.W. Perfect metamaterial absorber with dual bands. *PIER* **2010**, *108*, 37–49. [CrossRef]

12. Jamilan, S.; Azarmanesh, M.N.; Zarifi, D. Design and characterization of a dual-band metamaterial absorber based on destructive interferences. *PIER C* **2014**, *47*, 95–101. [CrossRef]

13. Dincer, F.; Karaaslan, M.; Unal, E.; Delihacioglu, K.; Sabah, C. Design of polarization and incident angle insensitive dual-band metamaterial absorber based on isotropic resonator. *PIER* **2014**, *144*, 123–132. [CrossRef]

14. Park, J.W.; Tuong, P.V.; Rhee, J.Y.; Kim, K.W.; Jang, W.H.; Choi, E.H.; Chen, L.Y.; Lee, Y.P. Multi-band metamaterial absorber based on the arrangement of donut-type resonators. *Opt. Express* **2013**, *21*, 9691–9702. [CrossRef] [PubMed]

15. Shen, X.; Cui, T.J.; Zhao, J.; Ma, H.F.; Jiang, W.X.; Li, H. Polarization-independent wide-angle triple-band metamaterial absorber. *Opt. Express* **2011**, *19*, 9401–9407. [CrossRef] [PubMed]

16. Mishra, N.; Choudhary, D.K.; Chowdhury, R.; Kumari, K.; Chaudhary, R.K. An investigation on compact ultra-thin triple band polarization independent metamaterial absorber for microwave frequency applications. *IEEE Access* **2017**, *5*, 4370–4376. [CrossRef]

17. Huang, L.; Chen, H. Multi-band and polarization insensitive metamaterial absorber. *PIER* **2011**, *113*, 103–110. [CrossRef]

18. Costanzo, S.; Venneri, F. Miniaturized fractal reflectarray element using fixed-size patch. *IEEE Antennas Wirel. Propag. Lett.* **2014**, *13*, 1437–1440. [CrossRef]

19. Costanzo, S.; Venneri, F.; Di Massa, G.; Borgia, A.; Costanzo, A.; Raffo, A. Fractal reflectarray antennas: State of art and new opportunities. *Int. J. Antennas Propag.* **2016**, *2016*, 17. [CrossRef]
20. Costa, F.; Genovesi, S.; Monorchio, A.; Manara, G. A circuit-based model for the interpretation of perfect metamaterial absorbers. *IEEE Trans. Antennas Propag.* **2013**, *61*, 1201–1209. [CrossRef]
21. Costanzo, S.; Di Massa, G. An integrated probe for phaseless near-field measurements. *Measurement* **2002**, *31*, 123–129. [CrossRef]
22. Costanzo, S.; Di Massa, G. Direct far-field computation from bi-polar near-field samples. *J. Electromagn. Waves Appl.* **2006**, *20*, 1137–1148. [CrossRef]

Article

Analytical Modeling of Metamaterial Differential Transmission Line Using Corrugated Ground Planes in High-Speed Printed Circuit Boards

Myunghoi Kim

Department of Electrical, Electronic, and Control Engineering, and the Institute for Information Technology Convergence, Hankyong National University, Anseong 17579, Korea; mhkim80@hknu.ac.kr

Received: 11 January 2019; Accepted: 1 March 2019; Published: 7 March 2019

Abstract: An analytical model for metamaterial differential transmission lines (MTM-DTLs) with a corrugated ground-plane electromagnetic bandgap (CGP-EBG) structure in high-speed printed circuit boards is proposed. The proposed model aims to efficiently and accurately predict the suppression of common-mode noise and differential signal transmission characteristics. Analytical expressions for the four-port impedance matrix of the CGP-EBG MTM-DTL are derived using coupled-line theory and a segmentation method. Converting the impedance matrix into mixed-mode scattering parameters enables obtaining common-mode noise suppression and differential signal transmission characteristics. The comprehensive evaluations of the CGP-EBG MTM-DTL using the proposed analytical model are also reported, which is validated by comparing mixed-mode scattering parameters S_{cc21} and S_{dd21} with those obtained from full-wave simulations and measurements. The proposed analytical model provides a drastic reduction of computation time and accurate results compared to full-wave simulation.

Keywords: common-mode noise; corrugated ground plane; differential signaling; electromagnetic bandgap; metamaterial; stepped impedance

---

1. Introduction

In high-speed printed circuit boards (PCBs) for digital communication systems, such as USB, HDMI, and PCI-Express technologies, electromagnetic interference during data transmission is a serious problem because the data rate of digital interfaces and the switching speed of processors continuously increase, but the voltage level continuously decreases. In recent high-speed PCB designs, simultaneous switching output noise in power delivery networks and near-field coupling noise in high-speed transmission lines notably undermine digital data transmission. To mitigate the electromagnetic interference effects, differential signaling is usually adopted. Given its balanced transmission, differential signaling is tolerant to both simultaneous switching output noise and near-field coupling noise. Hence, differential signaling is being increasingly adopted in high-speed PCBs by implementing a scheme using three conductor systems, namely positive signal line, negative signal line, and ground plane. Maintaining symmetry of those signal lines over the ground plane in high-speed PCBs is critical, because the noise robustness of differential signaling depends on the balanced scheme. However, it is difficult to prevent asymmetries in practical PCB designs. For instance, the unbalanced structures of meander delay lines, ball–grid–array escape routing, and asymmetric ground via configurations frequently occur during the practical design of high-speed PCBs. Moreover, current-strength mismatch of output driver circuits and trace-length mismatch in cables lead to unbalance in the differential scheme even if the PCB is symmetrical. Consequently, imbalance in differential signaling is inevitable in practical PCBs.

Unbalanced differential lines result in severe common-mode (CM) noise generation in high-speed PCBs [1–4]. Specifically, skewed signals in unbalanced differential lines generate wideband CM noise with high-order harmonics in the range of gigahertz and related electromagnetic interference problems. For intra-system electromagnetic interference, CM noise coupled to power delivery networks degrades the noise margin and timing budget of circuits mounted on the same board. Additionally, serious radio-frequency interference is produced when the CM noise flows into cables attached to PCBs. A practical example of radio-frequency interference induced by CM noise is presented in Reference [5], where experimental results show severe interference between a USB device and a 2.4-GHz wireless LAN device. Hence, CM noise caused by unbalanced differential structures should be mitigated in high-speed PCBs.

To suppress wideband and high-frequency CM noise in high-speed PCBs, metamaterial (MTM)-based techniques were proposed [6–11]. In Reference [6], differential transmission lines (DTLs), using a mushroom-type electromagnetic bandgap (EBG) structure with LTCC process technology, are presented. This type of EBG structure provides a wide stopband for even-mode propagation to suppress CM noise flowing along the differential lines, and the passband characteristics of odd-mode propagation ensure good differential-signal integrity during transmission. Then, CM noise suppression is predicted using dispersion analysis based on a lumped circuit model assuming a periodic structure, but prediction for differential signal transmission is not provided. In Reference [7], a complementary split ring resonator (CSRR) is employed to suppress CM noise of differential lines in high-speed PCBs. The CSRR etched in a ground plane forms an LC resonator to prevent even-mode propagation and behaves as an LC ladder network for odd-mode propagation. To predict selective CM noise suppression, a dispersion relationship is established from a lumped circuit model. However, the differential transmission characteristics are not estimated before performing full-wave simulations and measurements. As reported in References [8–11], a stepped impedance resonator is favorable for MTM-DTLs. In Reference [8], a stepped impedance resonator is implemented using various sizes for planar patches. A large patch and a narrow branch correspond to low and high characteristic impedances, respectively. CM noise suppression and differential signal quality are examined through full-wave simulations and experiments. In Reference [9], dual-type transmission lines are proposed to realize an MTM-DTLs. Impedance variations are implemented by alternating microstrip and strip lines, and a dispersion equation based on a periodic condition is derived, but it only predicts CM noise suppression. In References [10,11], a technique with ground planes vertically distributed in PCBs is proposed. The stepped impedance for odd-mode propagation is realized by variations of the vertical distances from differential lines to ground planes, representing a corrugated ground-plane electromagnetic bandgap (CGP-EBG) structure. Dispersion equations for odd-mode propagation are extracted to predict CM noise suppression only.

Therefore, recent research mainly focused on MTM-DTLs for CM noise suppression. It demonstrated wideband and high-frequency CM noise suppression with good differential signal integrity, simple implementation, and rigorous analysis based on dispersion relations. The key idea behind the development of MTM-DTLs is to provide structures handling different characteristics in the propagation modes of CM noise and differential-mode (DM) signal. To efficiently design and optimize MTM-DTLs, predicting and examining CM and DM characteristics of MTM-DTLs before their fabrication is needed. Although electromagnetic full-wave simulations can be employed, their prediction is computationally intensive in time and resources, thus being unsuitable for rapid design, testing, and prototyping of MTM-DTLs.

To overcome this problem, dispersion analysis with Floquet theory was adopted to estimate CM noise suppression. However, this analysis is limited to periodic structures, hindering the estimation of noise suppression characteristics when MTM-DTLs include a finite and small number of unit cells (UCs). Moreover, the periodicity of an MTM-DTL is not ensured in practical PCB applications. Likewise, dispersion analysis does not allow to accurately predict the suppression level, as its results only indicate whether a sufficient suppression will be provided in a stopband without retrieving a quantitative prediction. In addition, CM noise suppression characteristics in regions outside the stopband are not obtained, despite this information being critical in certain MTM-DTL applications. Moreover, the DM propagation associated with differential signaling characteristics was not estimated with a fast and simple approach in previous works. In Reference [12], the analytical method for the EBG structure employed in power delivery networks is presented. However, it is difficult to apply the method to DTLs because the segment model and recombined Z-parameter are limited to power delivery networks. Consequently, an efficient and accurate method to predict both CM noise and differential signaling characteristics of MTM-DTLs is still required for practical research and development of high-speed PCBs.

In this paper, an analytical model focusing on the MTM-DTL using a CGP-EBG structure is presented. In previous research [11], the CGP-EBG MTM-DTL was not fully characterized due to a limited method of CM noise prediction and a lack of estimation of differential transmission characteristics. Therefore, the research of the CGP-EBG MTM-DTL is extended by proposing an analytical model that efficiently provides rapid and accurate results for a nonperiodic array of CGP-EBG MTM-DTLs. The contribution of this paper is developing and verifying the analytical model, which simultaneously and quantitatively estimates the CM and DM propagation characteristics for the CGP-EBG MTM-DTL with a finite number of UCs.

2. Analytical Model of CGP-EBG MTM-DTL

2.1. Description of CGP-EBG MTM-DTL

In this section, a CGP-EBG MTM-DTL is briefly described, and the method for its analytical modeling is developed. Figure 1 shows a UC with its geometric parameters and the top and side views of the CGP-EBG MTM-DTL with a finite array size (three UCs). This CGP-EBG technique forms a stepped impedance resonator for even-mode propagation associated with CM noise and a constant impedance for odd-mode propagation associated with differential signal transmission by using the original distribution of ground planes. Basically, the characteristic impedances for odd- (Z_{oo}) and even-mode (Z_{oe}) propagations are determined by combining various geometric parameters of signal lines and ground plane. However, the CGP-EBG technique retrieves the desired Z_{oe} and Z_{oo} by only adjusting the vertical distance between signal lines and ground plane. The specific arrangement of ground patches, which are distributed in the different layers, provides varying Z_{oe} but the same Z_{oo} given in References [10,11]. The vertical distribution of ground planes efficiently achieves the decomposition of CM noise and differential signal propagation. The bandgap characteristics of the periodic CGP-EBG MTM-DTL were analyzed in References [10,11].

As shown in Figure 1, the UC of the CGP-EBG MTM-DTL consists of three parts, namely two DTLs with low Z_{oe} (LZ-DTL) and length $d_L/2$, and one DTL with high Z_{oe} (HZ-DTL) and length d_H. The vertical distance h_L between signal transmission lines and a ground patch for LZ-DTL is relatively short compared to h_H of HZ-DTL for obtaining low Z_{oe} for LZ-DTL and high Z_{oe} for HZ-DTL. Those ground planes are connected using plated-through-hole or blind vias for DC continuity. The via is separated from the center of the differential line by distance S_v.

2.2. Analytical Model

The proposed model is based on an analytical expression for the impedance matrix (Z-matrix) of LZ- and HZ-DTLs and a segmentation method for recombining the Z-matrices. The procedure of the proposed method mainly consists of two steps, namely segment modeling and segment recombination, as illustrated in Figure 2. During segment modeling, analytical expressions for the Z-parameters of HZ- and LZ-DTLs are extracted using coupled-line theory. Moreover, an analytical model of the via, which is used for reference plane transition between HZ- and LZ-DTLs, is derived.

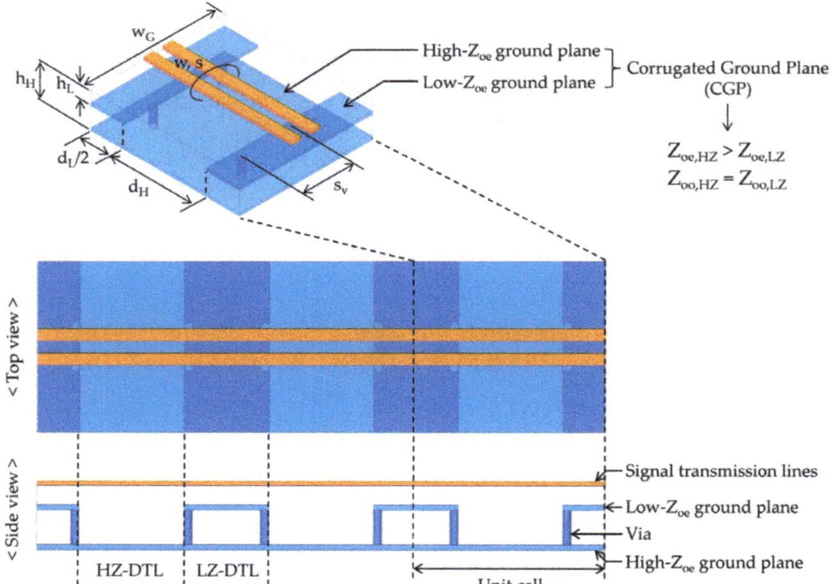

Figure 1. Schematic of a metamaterial differential transmission line (MTM-DTL) using a corrugated ground-plane electromagnetic bandgap (CGP-EBG) structure.

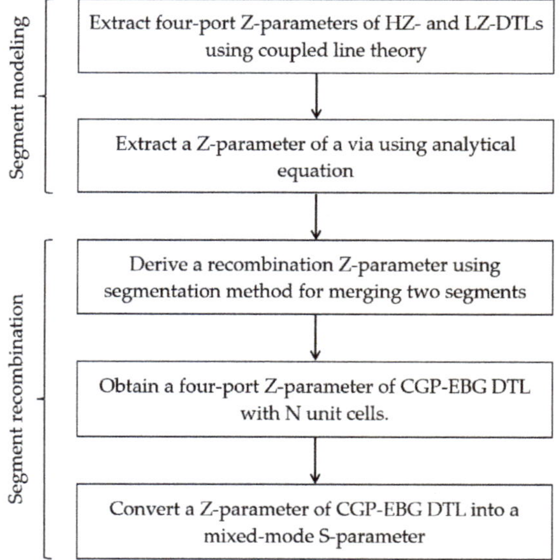

Figure 2. Procedure of proposed analytical method for predicting common-mode (CM) noise suppression and differential signal transmission of CGP-EBG MTM-DTLs.

The Z-parameters of the HZ- and LZ-DTLs are extracted using microwave theory for the four-port parallel-coupled lines that are represented in Figure 3a. The coupled line can be characterized by characteristic impedances of even and odd modes, a propagation constant, and a line length. In the parallel-coupled line for the HZ- and LZ-DTLs, the odd- and even-mode characteristic impedances at the i-th segment for recombination are denoted as $Z_{oo(i)}$ and $Z_{oe(i)}$, respectively, whereas the propagation constant and length are denoted as β_i and d_i, respectively. The voltage–current relationship for the parallel-coupled line of the i-th segment is given by

$$
\begin{pmatrix} V_{1,\mathrm{seg}(i)} \\ V_{2,\mathrm{seg}(i)} \\ V_{3,\mathrm{seg}(i)} \\ V_{4,\mathrm{seg}(i)} \end{pmatrix} = \begin{pmatrix} Z_{11,\mathrm{seg}(i)} & Z_{12,\mathrm{seg}(i)} & Z_{13,\mathrm{seg}(i)} & Z_{14,\mathrm{seg}(i)} \\ Z_{21,\mathrm{seg}(i)} & Z_{22,\mathrm{seg}(i)} & Z_{23,\mathrm{seg}(i)} & Z_{24,\mathrm{seg}(i)} \\ Z_{31,\mathrm{seg}(i)} & Z_{32,\mathrm{seg}(i)} & Z_{33,\mathrm{seg}(i)} & Z_{34,\mathrm{seg}(i)} \\ Z_{41,\mathrm{seg}(i)} & Z_{42,\mathrm{seg}(i)} & Z_{43,\mathrm{seg}(i)} & Z_{44,\mathrm{seg}(i)} \end{pmatrix} \begin{pmatrix} I_{1,\mathrm{seg}(i)} \\ I_{2,\mathrm{seg}(i)} \\ I_{3,\mathrm{seg}(i)} \\ I_{4,\mathrm{seg}(i)} \end{pmatrix} \tag{1}
$$

where

$$
Z_{11,\mathrm{seg}(i)} = Z_{11,\mathrm{seg}(i)} = Z_{11,\mathrm{seg}(i)} = Z_{11,\mathrm{seg}(i)} = -j/2\left(Z_{\mathrm{oe}(i)} + Z_{\mathrm{oo}(i)}\right)\cot(\beta_i d_i), \tag{2a}
$$

$$
Z_{12,\mathrm{seg}(i)} = Z_{21,\mathrm{seg}(i)} = Z_{34,\mathrm{seg}(i)} = Z_{43,\mathrm{seg}(i)} = -j/2\left(Z_{\mathrm{oe}(i)} - Z_{\mathrm{oo}(i)}\right)\cot(\beta_i d_i), \tag{2b}
$$

$$
Z_{13,\mathrm{seg}(i)} = Z_{31,\mathrm{seg}(i)} = Z_{24,\mathrm{seg}(i)} = Z_{42,\mathrm{seg}(i)} = -j/2\left(Z_{\mathrm{oe}(i)} - Z_{\mathrm{oo}(i)}\right)\csc(\beta_i d_i), \tag{2c}
$$

$$
Z_{14,\mathrm{seg}(i)} = Z_{41,\mathrm{seg}(i)} = Z_{23,\mathrm{seg}(i)} = Z_{32,\mathrm{seg}(i)} = -j/2\left(Z_{\mathrm{oe}(i)} + Z_{\mathrm{oo}(i)}\right)\csc(\beta_i d_i). \tag{2d}
$$

The analytical expressions for the Z-parameter of the i-th segment in Equation (2) are obtained from References [13,14]. The Z-parameter of the HZ- and LZ-DTLs are extracted by letting $(Z_{\mathrm{oe}(i)} = Z_{\mathrm{oe,H}}, Z_{\mathrm{oo}(i)} = Z_{\mathrm{oo,H}})$ and $(Z_{\mathrm{oe}(i)} = Z_{\mathrm{oe,L}}, Z_{\mathrm{oo}(i)} = Z_{\mathrm{oo,L}})$, respectively. The analytical expressions for the HZ- and LZ-DTLs at the i-th segment are obtained from Equations (2a)–(2d).

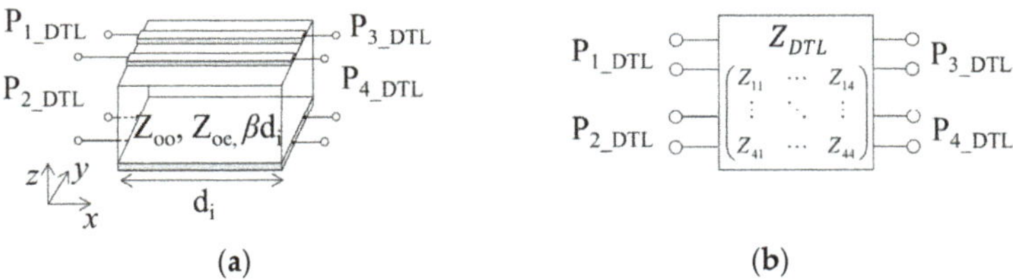

Figure 3. (a) Coupled line and (b) impedance parameter expression for a segment of CGP-EBG MTM-DTLs.

The physical transition from the HZ-DTL to the LZ-DTL is modeled to capture its inductive effect on return current. In particular, the return path for even-mode propagation is considered because the physical transition mainly contributes this propagation, as described in Reference [11]. The return path effect can be modeled as an effective partial inductance determined by the via diameter and length, the pitch between the via and the center of the signal transmission line, and the number of vias. In the proposed analytical method, only the via diameter and length are considered because they are the major parameters determining the effective inductance for the transitions. The inductive effect of the physical transition is analytically expressed as

$$
L_v = \begin{cases} \mu_0\dfrac{h_v}{2\pi}\left[\ln\left(\dfrac{2h_v}{r} + \sqrt{1+\left(\dfrac{2h_v}{r}\right)^2}\right) - \sqrt{1+\left(\dfrac{2h_v}{r}\right)^{-2}} + \left(\dfrac{2h_v}{r}\right)^{-1} + \dfrac14\right] & \text{for even mode,} \\ 0 \ \ \text{for odd mode,} \end{cases} \tag{3}
$$

where μ_0 is the permeability of free space, and h_v and r are the via length and radius, respectively, with $h_v = h_\mathrm{H} - h_\mathrm{L}$. The analytical expression for the via inductance in Equation (3) is extracted from Reference [15].

Next, segment recombination for the proposed model is applied. In this step, a technique is presented for merging the Z-matrices of HZ- and LZ-DTLs connected to each other through a via inductance to obtain the Z-matrix of a finite array of CGP-EBG MTM-DTLs. Segment recombination consists of three parts, namely derivation of recombined Z-matrix, iterative recombination, and parameter conversion. To extract the recombined Z-matrix, the block diagram of the segmentation method is illustrated in Figure 4. Two segments are connected through via inductances. For the sake of generality, the segments are denoted as the i-th and $(i + 1)$-th segments. Each segment includes four ports, P_{1_seg} to P_{4_seg}. The Z-parameters of the segments are $Z_{seg(i)}$ and $Z_{seg(i+1)}$. Port P_{3_seg} of the i-th segment is connected to port P_{1_seg} of the $(i + 1)$-th segment through the via with inductance L_v. Similarly, the other ports are connected through the via. Then, the recombined segment is conformed of Z-parameter and ports represented as $Z_{rec(i)}$, $P_{1\_rec(i)}$, $P_{2\_rec(i)}$, $P_{3\_rec(i)}$, and $P_{4\_rec(i)}$. Ports $P_{1\_rec(i)}$ and $P_{2\_rec(i)}$ of the recombined segment are associated with ports $P_{1\_seg(i)}$ and $P_{2\_seg(i)}$ of the i-th segment, whereas ports P_{3_rec} and P_{4_rec} of the recombined segment are associated with ports $P_{1\_seg(i+1)}$ and $P_{2\_seg(i+1)}$ of the $(i + 1)$-th segment. Using the segmentation method in Reference [16], recombined matrix $Z_{rec(i)}$ is extracted in terms of $Z_{seg(i)}$, $Z_{seg(i+1)}$, and L_v with operator $\oplus$ representing Z-parameter recombination.

$$
\begin{aligned}
Z_{rec(i)} &= Z_{seg(i)} \oplus Z_{via} \oplus Z_{seg(i+1)} \\
&= Z_{pp} + (Z_{pq} - Z_{pr})(Z_{qq} - Z_{qr} - Z_{rq} + Z_{rr} + Z_{via})^{-1}(Z_{rp} - Z_{qp}),
\end{aligned}
\tag{4}
$$

where

$$
Z_{pp} = \begin{pmatrix} Z_{11,seg(i)} & Z_{12,seg(i)} & 0 & 0 \\ Z_{21,seg(i)} & Z_{22,seg(i)} & 0 & 0 \\ 0 & 0 & Z_{33,seg(i+1)} & Z_{34,seg(i+1)} \\ 0 & 0 & Z_{43,seg(i+1)} & Z_{44,seg(i+1)} \end{pmatrix},
\tag{5a}
$$

$$
Z_{pq} = \begin{pmatrix} Z_{13,seg(i)} & Z_{14,seg(i)} \\ Z_{23,seg(i)} & Z_{24,seg(i)} \\ 0 & 0 \\ 0 & 0 \end{pmatrix},
\tag{5b}
$$

$$
Z_{pr} = \begin{pmatrix} 0 & 0 \\ 0 & 0 \\ Z_{31,seg(i+1)} & Z_{32,seg(i+1)}0 \\ Z_{41,seg(i+1)} & Z_{42,seg(i+1)} \end{pmatrix},
\tag{5c}
$$

$$
Z_{qq} = \begin{pmatrix} Z_{33,seg(i)} & Z_{34,seg(i)} \\ Z_{43,seg(i)} & Z_{44,seg(i)} \end{pmatrix},
\tag{5d}
$$

$$
Z_{qr} = Z_{rq} = \begin{pmatrix} 0 & 0 \\ 0 & 0 \end{pmatrix},
\tag{5e}
$$

$$
Z_{rr} = \begin{pmatrix} Z_{11,seg(i+1)} & Z_{12,seg(i+1)} \\ Z_{21,seg(i+1)} & Z_{22,seg(i+1)} \end{pmatrix},
\tag{5f}
$$

$$
Z_{via} = \begin{pmatrix} j\omega L_v & 0 \\ 0 & j\omega L_v \end{pmatrix},
\tag{5g}
$$

$$
Z_{rp} = \begin{pmatrix} 0 & 0 & Z_{13,seg(i+1)} & Z_{14,seg(i+1)} \\ 0 & 0 & Z_{23,seg(i+1)} & Z_{24,seg(i+1)} \end{pmatrix},
\tag{5h}
$$

$$
Z_{qp} = \begin{pmatrix} Z_{31,seg(i)} & Z_{32,seg(i)} & 0 & 0 \\ Z_{41,seg(i)} & Z_{42,seg(i)} & 0 & 0 \end{pmatrix}.
\tag{5i}
$$

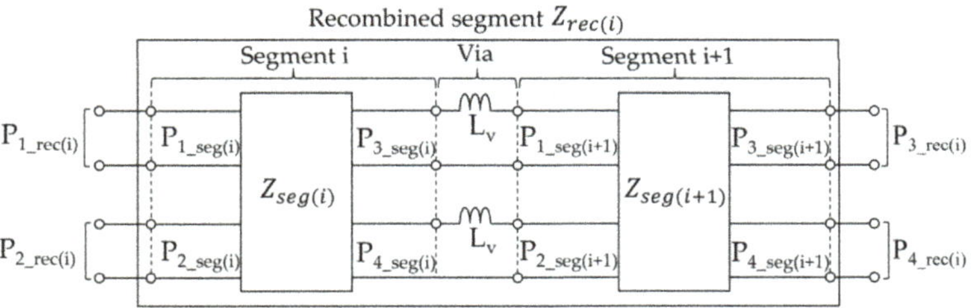

Figure 4. Block diagram to derive recombined Z-parameter using segmentation method.

In the next part, recombination of the Z-parameter is iteratively applied. Suppose that a CGP-EBG MTM-DTL containing N UCs is given. The original problem of the MTM-DTL is replaced with subdomain problems constituting the N UCs further divided into HZ-DTLs, LZ-DTLs, and via transitions. The solutions of the subdomains are obtained using the expressions in Equations (1)–(3). Applying the recombined Z-parameter method iteratively, the subdomain solutions are updated to finally obtain the Z-parameter of the CGP-EBG MTM-DTLs with N UCs as

$$Z_{\text{rec}(1)} = Z_{\text{DTL}}(Z_{\text{oe,L}}, Z_{\text{oo}}, d_{\text{L}}/2) \oplus Z_{\text{via}} \oplus Z_{\text{DTL}}(Z_{\text{oe,H}}, Z_{\text{oo}}, d_{\text{H}}), \tag{6a}$$

$$Z_{\text{rec}(2)} = Z_{\text{rec}(1)} \oplus Z_{\text{via}} \oplus Z_{\text{DTL}}(Z_{\text{oe,L}}, Z_{\text{oo}}, d_{\text{L}}), \tag{6b}$$

$$Z_{\text{rec}(2(N-1))} = Z_{\text{rec}(2(N-1)-1)} \oplus Z_{\text{via}} \oplus Z_{\text{DTL}}(Z_{\text{oe,H}}, Z_{\text{oo}}, d_{\text{H}}), \tag{6c}$$

$$Z_{\text{MTM−DTL}} = Z_{\text{rec}(2(N-1)+1)} \oplus Z_{\text{via}} \oplus Z_{\text{DTL}}\left(Z_{\text{oe,L}}, Z_{\text{oo}}, \frac{d_{\text{L}}}{2}\right), \tag{6d}$$

where $Z_{\text{DTL}}(Z_{\text{oe,L}}, Z_{\text{oo}}, d_{\text{L}}/2)$ denotes a four-port Z-parameter of the LZ-DTL with the length being half of d_{L}. The first and last segments of the CGP-EBG MTM-DTL, presented herein, are supposed to be the LZ-DTLs with length $d_{\text{L}}/2$. Still, note that the proposed analytical model for the MTM DTL is not limited to this configuration. From the four-port Z-parameter of the CGP-EBG MTM-DTL, the S-parameter is given as

$$S_{\text{MTM−DTL}} = (Z_{\text{MTM−DTL}} + Z_0 E)^{-1}(Z_{\text{MTM−DTL}} - Z_0 E), \tag{7}$$

where Z_0 is the reference characteristic impedance, commonly taken as 50 Ω, and E is the identity matrix. The CM and differential characteristics are finally obtained using mixed-mode S-parameter theory [17]. Only the analytical expressions for S_{cc21} and S_{dd21} are presented due to the focus on CM noise suppression and differential transmission characteristics.

$$S_{\text{cc21}} = \frac{1}{2}(S_{31,\text{MTM−DTL}} + S_{41,\text{MTM−DTL}} + S_{32,\text{MTM−DTL}} + S_{42,\text{MTM−DTL}}), \tag{8}$$

$$S_{\text{dd21}} = \frac{1}{2}(S_{31,\text{MTM−DTL}} - S_{41,\text{MTM−DTL}} - S_{32,\text{MTM−DTL}} + S_{42,\text{MTM−DTL}}). \tag{9}$$

3. Results

To verify the proposed analytical model of CGP-EBG MTM-DTLs, it is compared to full-wave simulations based on the finite element method (FEM) [18] and measurements. Comprehensive validations were performed considering various cases. Firstly, the CM characteristics of the CGP-EBG MTM-DTLs between the proposed model and full-wave simulations are compared. For the full-wave simulation model, geometrical parameters d_{H}, h_{H}, d_{L}, h_{L}, w, s, w_{G}, and s_{v} are set to 10, 1.0, 10, 0.08, 0.1, 0.1, 10, and 1.3 mm, respectively. The values were determined considering a commercial PCB process. FR-4 and copper (35 μm thick) were used as a dielectric material and conductor, respectively.

The physical dimensions of the CGP-EBG MTM-DTL present $Z_{oe,H}$, $Z_{oo,H}$, $Z_{oe,L}$, and $Z_{oo,L}$ of 218, 66.3, 66, and 52 Ω, respectively. The dimensions and parameters for this validation are listed in Table 1.

The FEM simulation model for the CGP-EBG MTM-DTLs including three unit cells and their mesh generation result are shown in Figure 5. The waveguide ports are adopted for the excitation of the CM and DM waves at ports 1, 2, 3, and 4 of the differential signal lines. The perfect magnetic conductor and radiation boundaries are assigned at the sides and top of the vacuum box, which is shown as a red box in the Figure 5a. The dielectric constant and loss tangent of FR-4 are 4.4 and 0.02, respectively. The via in the FEM model is modeled using a polyhedron with 12 segments. Its radius and height are 0.2 and 0.9 mm, respectively. The meshes are generated with the solution frequency of 10 GHz, which is the maximum frequency of interest. As can be seen in Figure 5b, most meshes are placed in differential signal lines and ground plane transition.

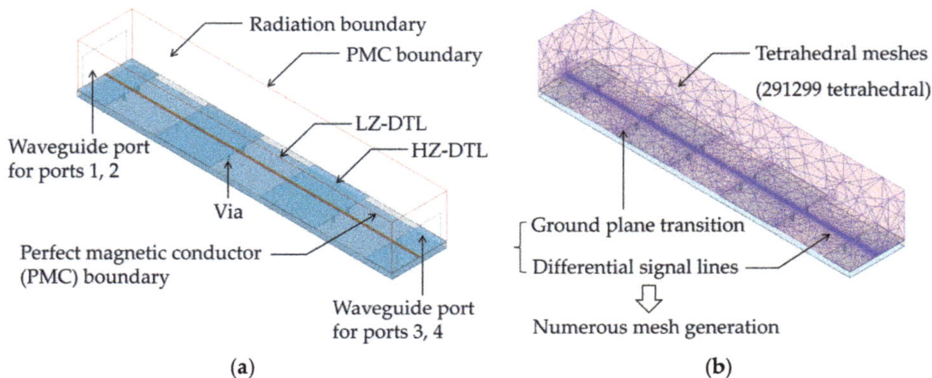

Figure 5. (a) Finite element method (FEM) simulation model, and (b) mesh generation result of CGP-EBG MTM-DTLs with three unit cells.

Figure 6 depicts the eight curves of parameter S_{cc21} from the CGP-EBG MTM-DTLs considering one to four UCs, using the proposed analytical model (solid lines) and full-wave simulations (dashed lines). The proposed model suitably agrees with the full-wave simulations in the four cases. From the proposed model, the minimum values of CM noise suppression from one to four UCs are −6.1, −15.6, −25.9, and −36.3 dB, respectively. Overall, CM noise suppression improves as the number of UCs increases.

Table 1. Parameters for verification of the proposed analytical model. DTL—differential transmission line; HZ—high Z_{oe}; LZ—low Z_{oe}; UC—unit cell; CM—common mode; DM—differential mode.

Parameter	Figure	HZ-DTL				LZ-DTL				No. of UCs
		$Z_{oe,H}$	$Z_{oo,H}$	d_H	h_H	$Z_{oe,L}$	$Z_{oo,L}$	d_L	h_L	
No. of UCs effect on CM, DM	6 (CM) 8 (DM)	218	66.3	10	1.0	66	52	10	0.08	1
		218	66.3	10	1.0	66	52	10	0.08	2
		218	66.3	10	1.0	66	52	10	0.08	3
		218	66.3	10	1.0	66	52	10	0.08	4
Z_{oe} effect on CM	7	113	66.3	10	0.2	66	52	10	0.08	4
		171	66.3	10	0.5	66	52	10	0.08	4
		218	66.3	10	1.0	66	52	10	0.08	4

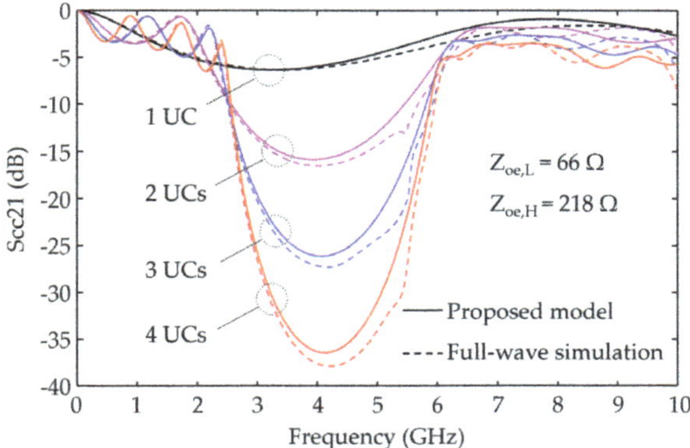

Figure 6. CM noise suppression with varying number of unit cells (UCs) for model verification.

Examining the suppression levels where the suppression bandwidth of 3 GHz is ensured, the values for two to four UCs are -10.1, -14.1, and -18.6 dB, respectively. The value of 3 GHz was selected because it corresponds to the suppression bandwidth predicted by the dispersion analysis based on Floquet theory. The high and low cut-off frequencies from Floquet theory were obtained as 2.7 and 5.7 GHz, respectively. The periodic analysis only estimates the cut-off frequencies, but not the suppression level. As seen in Figure 6, the suppression level corresponding to Floquet theory notably varies according to the number of UCs. In the practical use of MTM-DTLs for high-speed PCBs, a periodic condition is not commonly ensured, thus requiring the development of approaches such as the proposed model considering MTM DTLs with a finite and small number of UCs.

Remarkably, the proposed analytical model achieves a drastic reduction in computation time compared to the full-wave simulation. For instance, the time for determining S_{cc21} of the CGP-EBG MTM-DTL with four UCs using the proposed model was 0.3 s, whereas that using the full-wave simulation was 18,257 s. Hence, the proposed model substantially reduces the computation time while providing a suitable accuracy compared to the full-wave simulation. The time reduction results are listed in Table 2.

Table 2. Computation time for estimating CM and differential transmission characteristics of the corrugated ground-plane electromagnetic bandgap (CGP-EBG) metamaterial (MTM)-DTL.

	1 UC	2 UCs	3 UCs	4 UCs
Proposed model	0.13 s	0.19 s	0.24 s	0.32 s
Full-wave simulation	120 s	1830 s	5903 s	18,257 s

Computation platform: Intel Xeon processor (3.2 GHz), 512 GB RAM (E5-2667 v4 @3.20 GHz, Intel, Santa Clara, CA, USA).

The effect of Z_{oe} on parameter S_{cc21} is further examined by comparing the proposed analytical model and full-wave simulations, as shown in Figure 7. For the CGP-EBG MTM-DTL with four UCs, $Z_{oe,H}$ changes to 113, 171, and 218 Ω by adjusting h_H to 0.2, 0.5, and 1.0 mm, respectively. The corresponding changes in the ratio of $Z_{oe,H}$ to $Z_{oe,L}$ are approximately 1.7, 2.6, and 3.3. The low and high cut-off frequencies with a suppression level of -10 dB were also investigated. The corresponding (high, low) cut-off frequencies for $Z_{oe,H}$ of 113, 171, and 218 Ω are (3.56, 4.85 GHz), (2.94, 5.72 GHz), and (2.65, 6.09 GHz), respectively. Hence, the suppression bandwidth and level improve as the ratio of $Z_{oe,H}$ to $Z_{oe,L}$ increases. Again, the results of the proposed analytical model are consistent with those of the full-wave simulation.

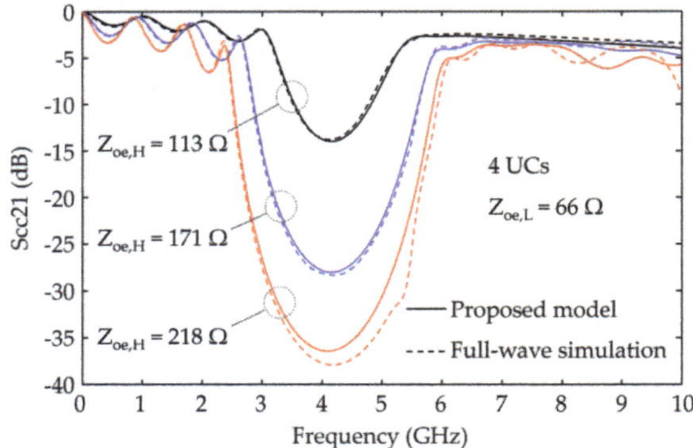

Figure 7. Effect of even-mode characteristic impedance (Z_{oe}) on CM noise suppression for model verification.

In addition to CM noise suppression, the differential transmission characteristics of the CGP-EBG MTM-DTLs were investigated. Figure 8 shows parameter S_{dd21} for differential transmission using the proposed model and full-wave simulations. The design parameters and dimensions are those listed in Table 1. The values of $Z_{oo,L}$ and $Z_{oo,H}$ associated with differential transmission are 52 and 66.3 Ω, respectively. For ideal differential characteristics, the same values of Z_{oo} between the HZ- and LZ-DTLs are preferred. However, this condition is commonly limited by the design rules of commercial PCB processes, thus making it difficult to avoid different Z_{oo} values for HZ- and LZ-DTLs in practical high-speed PCBs. The effect of the Z_{oo} difference on the differential characteristics is shown in Figure 8. The small difference in the Z_{oo} values between HZ- and LZ-DTLs degrades differential transmission characteristics. This effect increases with the number of UCs. This phenomenon can be inferred considering the theory of a stepped impedance resonator. However, it is important to obtain the exact degradation of parameter S_{dd21} for measuring and quantitating its impact on differential signal transmission.

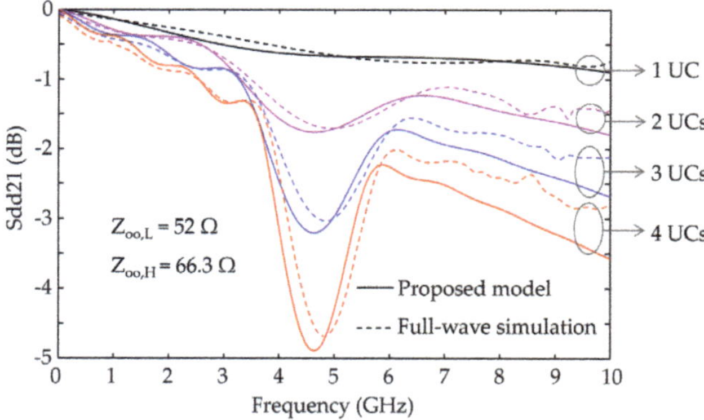

Figure 8. Differential signal transmission characteristics with varying number of UCs for model verification.

The accuracy and efficiency of the proposed analytical model for CGP-EBG MTM-DTLs are verified from comparisons to full-wave simulations based on FEM. The CM noise and differential signal transmission results of the proposed model suitably agree with those of full-wave simulations, but the computation time for obtaining the characteristics of the nonperiodic array of MTM-DTLs was substantially reduced using the proposed four-port analytical model.

To further validate the proposed model of the CGP-EBG MTM-DTL, the correlations between the proposed model, full-wave simulations, and measurements are examined using the fabricated PCB pattern of the CGP-EBG MTM-DTL. The fabricated PCB pattern and the measurement set-up for S_{cc21} and S_{dd21} are shown in Figure 9. The low-cost PCB process employs the FR-4 dielectric and copper metal layers. The dielectric constant and loss tangent of the FR-4 substrate are 4.4 and 0.02, respectively. The PCB pattern contains four HZ-DTLs and three LZ-DTLs. The geometric dimensions are shown in Figure 9a. To obtain the S_{cc21} and S_{dd21}, four-port S-parameters of the PCB pattern are measured using a vector network analyzer and high-frequency microprobes. Figure 10 depicts the comparison of S_{cc21} and S_{dd21} between the proposed model, full-wave simulations, and measurements. The proposed model agrees well with the measurements. As can be seen in the results, CM noise is successfully suppressed, while good differential data transmission is achieved. In the S_{cc21} of the measurements, discrepancy is observed around the frequency of 5 GHz. This defect is caused by PCB manufacturing tolerances because it is not observed in the full-wave simulations.

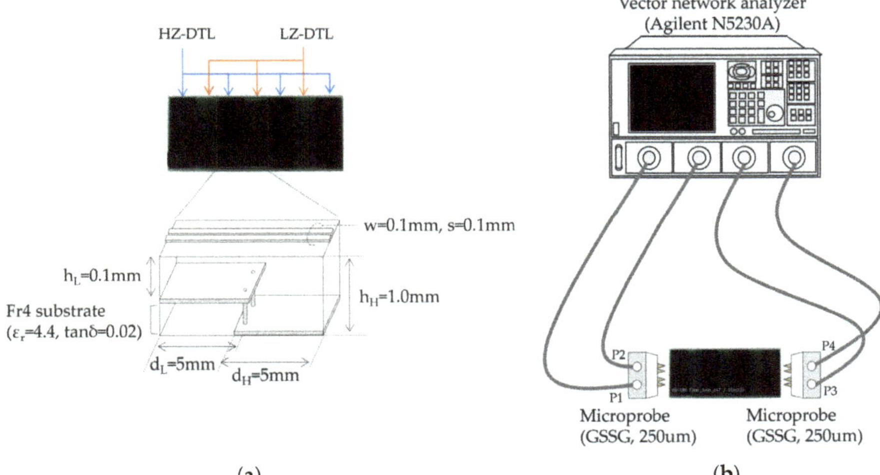

Figure 9. (**a**) Fabricated CGP-EBG MTM-DTL, and (**b**) measurement set-up for S-parameters.

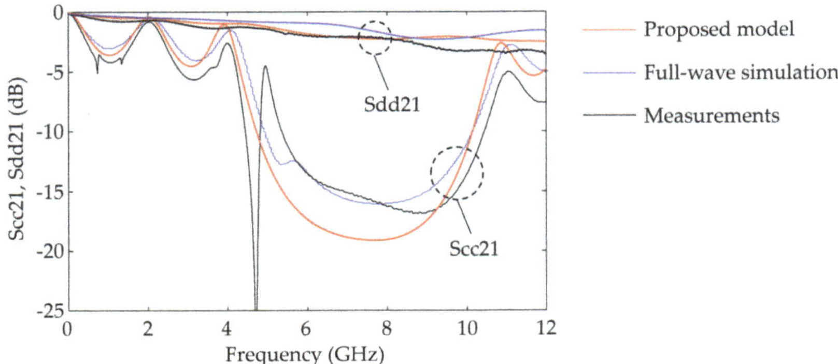

Figure 10. Comparison of S_{cc21} and S_{dd21} between proposed model, full-wave simulations, and measurements.

4. Conclusions

An efficient and accurate approach to evaluate CGP-EBG MTM-DTLs using an analytical model and segmentation method was proposed. The four-port model of the CGP-EBG segments was extracted using coupled-line theory and an analytical via model. For a finite array of CGP-EBG MTM-DTLs, the method of Z-parameter recombination was presented. The proposed analytical model was thoroughly validated by comparisons with full-wave simulations based on FEM. The CM noise suppression and differential signal transmission results of the CGP-EBG MTM-DTL variants were consistent between the proposed model and full-wave simulations. However, the computation time for estimating the characteristics of the CGP-EBG MTM-DTL was drastically reduced when using the proposed model. Moreover, the proposed model can be easily combined with other types of circuit and electromagnetic models, thus enabling system-level simulations for high-speed PCB applications. Overall, the assessment of CGP-EBG MTM-DTLs in early design stages and simulation-based verification can be efficiently and accurately conducted using the proposed analytical model. In this paper, the CGP-EBG MTM-DTL for one pair of differential lines was explored. For further research, the CGP-EBG MTM-DTL and its analytical model of multiple pairs of differential lines for emerging technologies of high-speed differential signaling can be examined.

Author Contributions: The author conceived and designed the experiments, analyzed the characteristics, performed the simulations and experiments, and wrote the paper.

Funding: This work was supported by the National Research Foundation of Korea (NRF) grant funded by the Korea government (Ministry of Science, ICT & Future Planning) (NRF-2016R1C1B1007123).

Conflicts of Interest: The author declares no conflicts of interest.

References

1. Fan, J.; Ye, X.; Kim, J.; Archambeault, B.; Orlandi, A. Signal integrity design for high-speed digital circuits: Progress and directions. *IEEE Trans. Electromagn. Compat.* **2010**, *52*, 392–400.
2. Shiue, G.; Yeh, C.; Liao, H.Y.; Huang, P. Significant Reduction of Common-Mode Noise in Weakly Coupled Differential Serpentine Delay Microstrip Lines Using Different- Layer-Routing-Turned Traces. *IEEE Trans. Compon. Packag. Manuf. Technol.* **2014**, *4*, 1671–1683. [CrossRef]
3. Lin, D.; Huang, C.; Ke, H. Using Stepped-Impedance Lines for Common-Mode Noise Reduction on Bended Coupled Transmission Lines. *IEEE Trans. Compon. Packag. Manuf. Technol.* **2016**, *6*, 757–766. [CrossRef]
4. Archambeault, B.; Diepenbrock, J.C.; Connor, S. EMI Emissions from mismatches in High Speed Differential Signal Traces and Cables. In Proceedings of the IEEE International Symposium on Electromagnetic Compatibility, Honolulu, HI, USA, 3–5 June 2007; pp. 1–6.
5. USB 3.0 Radio Frequency Interference Impact on 2.4 GHz Wireless Devices, White Paper. Intel Corporation, 2012. Available online: https://www.intel.com/content/www/us/en/io/universal-serial-bus/usb3-frequency-interference-paper.html (accessed on 11 January 2019).
6. Tsai, C.-H.; Wu, T.-L. A broadband and miniaturized common-mode filter for gigahertz differential signals based on negative-permittivity metamaterials. *IEEE Trans. Microw. Theory Technol.* **2010**, *58*, 195–202. [CrossRef]
7. Naqui, J.; Fernandez-Prieto, A.; Duran-Sindreu, M.; Mesa, F.; Martel, J.; Medina, F.; Martin, F. Common-Mode Suppression in Microstrip Differential Lines by Means of Complementary Split Ring Resonators: Theory and Applications. *IEEE Trans. Microw. Theory Tech.* **2012**, *60*, 3023–3034. [CrossRef]
8. Varner, M.A.; de Paulis, F.; Orlandi, A.; Connor, S.; Cracraft, M.; Archambeault, B.; Nisanci, M.H.; Di Febo, D. Removable EBG-Based Common-Mode Filter for High-Speed Signaling: Experimental Validation of Prototype Design. *IEEE Trans. Microw. Theory Technol.* **2012**, *57*, 672–679. [CrossRef]
9. Engin, A.E.; Modi, N.; Oomori, H. Stepped-Impedance Common-Mode Filter for Differential Lines Enhanced with Resonant Planes. *IEEE Trans. Electromagn. Compat.* **2018**, *99*, 1–8. [CrossRef]
10. Kim, M.; Kim, S.; Bae, B.; Cho, J.; Kim, J.; Kim, J.; Ahn, D.S. Application of VSI-EBG structure to high-speed differential signals for wideband suppression of common-mode noise. *ETRI J.* **2013**, *35*, 827–837. [CrossRef]

11. Kim, M. Periodically Corrugated Reference Planes for Common-Mode Noise Suppression in High-Speed Differential Signals. *IEEE Trans. Electromagn. Compat.* **2016**, *58*, 619–622. [CrossRef]
12. Kim, M. Unit-Cell-Based Domain Decomposition Method for Efficient Simulation of a Truncated Electromagnetic Bandgap Structure in High-Speed PCBs. *Electronics* **2018**, *7*, 201. [CrossRef]
13. Jones, E.M.T.; Bolljahn, J.T. Coupled strip transmission line filters and directional couplers. *IRE Trans. Microw. Theory Technol.* **1956**, *4*, 78–81. [CrossRef]
14. Jensen, T.; Zhurbenko, V.; Krozer, V.; Meincke, P. Coupled Transmission Lines as Impedance Transformer. *IEEE Trans. Microw. Theory Tech.* **2007**, *55*, 2957–2965. [CrossRef]
15. Wang, C.; Shiue, G.; Guo, W.; Wu, R. A Systematic Design to Suppress Wideband Ground Bounce Noise in High-Speed Circuits by Electromagnetic-Bandgap-Enhanced Split Powers. *IEEE Trans. Microw. Theory Tech.* **2006**, *54*, 4209–4217. [CrossRef]
16. Okoshi, T. *Planar Circuits for Microwaves and Lightwaves*; Springer: Munich, Germany, 1985.
17. Bockelman, D.; Einsenstadt, W.R. Combined differential and common-mode scattering parameters: Theory and simulation. *IEEE Trans. Microw. Theory Tech.* **1995**, *43*, 1530–1539. [CrossRef]
18. ANSYS HFSS, ANSYS, Inc. Available online: http://www.ansys.com (accessed on 11 January 2019).

Article

Two-Bit Terahertz Encoder Realized by Graphene-Based Metamaterials

Shan Yin [1,2], Xintong Shi [1], Wei Huang [1,*], Wentao Zhang [1,*], Fangrong Hu [1], Zujun Qin [1] and Xianming Xiong [1]

1   Guangxi Key Laboratory of Optoelectronic Information Processing, Guilin University of Electronic Technology, Guilin 541004, China; syin@guet.edu.cn (S.Y.); xt806@outlook.com (X.S.); hufangrong@sina.com (F.H.); qinzj@guet.edu.cn (Z.Q.); xmxiong@guet.edu.cn (X.X.)
2   Guangxi Colleges and Universities Key Laboratory of Complex System Optimization and Big Data Processing, Yulin Normal University, Yulin 537000, China
*   Correspondence: weihuang@guet.edu.cn (W.H.); zhangwentao@guet.edu.cn (W.Z.)

Received: 5 November 2019; Accepted: 7 December 2019; Published: 12 December 2019

Abstract: Terahertz (THz) technologies have achieved great progress in the past few decades. Developing active devices to control the THz waves is the frontier of THz applications. In this paper, a new scheme of two-bit THz encoder is proposed. Different from the present THz modulators whose spectra at different bands are varied simultaneously, our encoder can realize the individually efficient modulation of every channel. The encoder comprises the double-sided graphene-based metamaterials, in which the graphene structures on each side are connected to the external electrodes individually. The well-designed metamaterials on the front and back sides determine the resonances at two different bands (0.20 THz and 0.33 THz) separately. Through simulating the performance of this device by changing the conductivities of the graphene on each side independently, we demonstrate two-bit encoding realized by the dual-band modulation of transmission amplitude with electronic control, and the modulation depth can reach as high as 79.6%. Our encoder can promote the development of multifunctional and integrated devices, such as frequency division multiplexers and logical circuitry, which will contribute to THz communications.

Keywords: terahertz metamaterials; graphene; encoder; active control

1. Introduction

Terahertz (THz) radiation is generally referred to the frequency range of 0.1 to 10 THz, which is located between infrared and microwave bands in the electromagnetic spectrum. Due to its unique spectral characterization of the rotational and vibrational resonances, and the thermal emission lines of simple molecules, THz technologies have been an attractive research field, with applications for physics, medicine, biochemistry, astronomy, manufacturing, space and defense industries [1]. In the past few decades, THz technologies achieved great progress, especially in imaging, biological and medical applications, THz time-domain spectroscopy (THz-TDS) and non-destructive testing [2].

Taking advantage of ultrafast optics, the development of THz sources and detectors is much more mature. By contrast, the functional devices to manipulate THz waves are still been explored. A remarkable method to control THz waves is employing metamaterials. Metamaterials are usually referred to the elaborately arranged subwavelength structures [3], which can manipulate electromagnetic waves artificially. Based on this concept, researchers designed many on-demand functional devices, including flat lenses [4], polarization converters [5], filters [6], sensors [7], waveguides [8], etc. The previous works mainly employed passive devices, which face the drawback of single function and limited application scenarios due to the fixed established structures. Active devices can enrich their own functionalities by changing the material characteristics via controlling the applied physical quantity, like pump power of laser [9], voltage or current [10,11], magnetic forces or thermal [12], etc. Graphene is a competitive two-dimensional material for active control due to its large conductivity range and ultrafast response [13,14]. Graphene-based metamaterials used as waveguides [15], switches [16] and sensors [17] have shown their potential in developing optical devices. Meanwhile, the THz devices integrated with graphene emerged the probability in active control: Li et al. experimentally demonstrated an active diode for the THz waves consisting of a graphene–silicon hybrid film [18], and Kindness et al. achieved active resonance frequency tuning of a THz metamaterial by integrating metal-coupled resonator arrays with electrically tunable graphene [19]. Active devices are gradually being brought to the forefront of THz applications.

Another potential application of THz technology is THz communications. Compared to the microwave circuits, THz devices have the advantages of miniaturization, wideband and agility [20,21]. Since THz communications may play an important role in next-generation communications, THz coding technology will be of great significance. However, as one of the key components, THz encoder, which can compile information on THz waves, is rarely studied. Different from the multiband modulator whose spectra at different bands are varied simultaneously, an encoder needs to realize the individual control of every channel. Though there are many researches about graphene-based metamaterials whose transmittance can be governed by voltage [16,19,22], the independent control of different bands is hardly demonstrated. Currently, most THz modulators are made of in-plane metamaterials [23,24], where the problems caused by the interaction between different channels are intractable. One is the intersection of different electrodes in wiring, which will increase the complexity in fabrication. Another is the spatial arrangement of the metamaterials. If different resonators are too close, their coupling cannot be ignored; if they are too distant, the efficiency of the modulation will be reduced. So, developing an encoder is not a simple superposition of different modulators in one device. To solve those problems, we design the THz encoder by applying dual-sided hybrid metamaterials, which can separately control the voltages (front and back sides of our device) to achieve different coding function. Compared with the present THz modulators, our encoder can not only maximally enhance the coding number in the single chip, but also guarantee the modulation efficiency.

In this paper, we propose a two-bit THz encoder based on hybrid metamaterials, which can independently modulate two different bands with high efficiency. The encoder consists of the double-sided hybrid metamaterials patched with graphene deposited on the silicon substrate, where the graphene structures on each side are connected to the external electrodes individually. Utilizing CST (Computer Simulation Technology) microwave studio, we simulate the transmission spectra of the device, and we demonstrate two-bit encoding based on the dual-band independent switches by tuning the Fermi level of the graphene with voltage control. Using the double-sided metamaterials, we successfully eliminate the coupling between different bands, and meanwhile, the modulation depth can reach as high as 79.6%. Our encoder, with simple structure and easy operation, may contribute to the promotion of multifunctional and integrated devices, such as frequency division multiplexers and logical circuitry.

2. Device Design and Simulation

Our THz encoder was composed of hybrid graphene–metal–silica metamaterials laying on the front and back surfaces of the silicon substrate, respectively. We put two kinds of hybrid metamaterial-based resonators (chosen as classical split ring resonators, SRRs) on each side of the device, respectively, which determined the working bands of the encoder. Resonators on each surface were connected to the external electrodes individually. Therefore, we could separately control the dual-band transmission amplitudes via modulating the voltage applied on each side hybrid metamaterial.

The hybrid metamaterials were made of periodic unit cells diagramed in Figure 1a. The metallic structures functioned as the electromagnetic resonators, the graphene strips linked with electrodes worked as the controller, and the silica structures sandwiched between the metallic and graphene layers were to isolate the graphene strips and the silicon substrate [14]. The metallic structures on both surfaces were SRRs, whose shapes were square split rings on the front side and circle split rings on the other side. The closed ring-shaped silica structures were underneath the metallic layer, which were attached with strips along y axis on both surfaces. The single-layer graphene strips paralleled to y axis patch on the top of the metallic and silica structures of both side metamaterials.

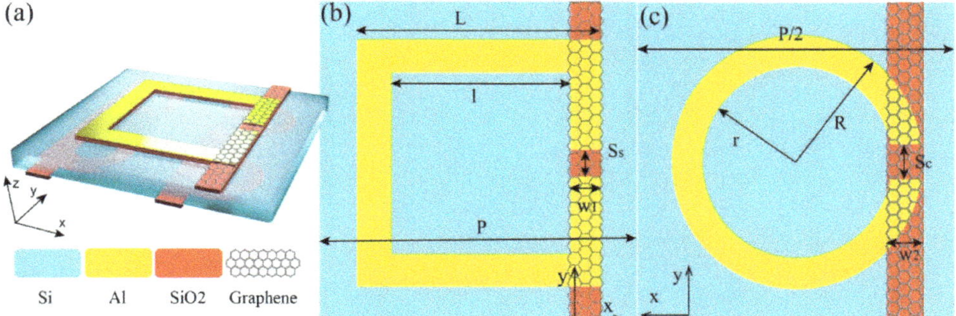

Figure 1. Structures of the unit cell: (**a**) The schematic of the unit cell, (**b**) the front and (**c**) the back views of the unit cell.

The key indicator for an encoder is the modulation depth. We have optimized and determined the geometrical parameters of two resonators by comparing their performance in modulation depth. As shown in Figure 1, the dimensions of our encoder were as follows: For the square split rings shown in Figure 1b, outer and inner lengths of the arms were $L = 130\ \mu m$ and $l = 100\ \mu m$, respectively, gap of square split ring $S_s = 10\ \mu m$, width of silica (graphene) strip $w_1 = 15\ \mu m$, and the period of unit cell $P = 180\ \mu m$; for the circle split rings shown in Figure 1c, outer and inner radii of the ring were $R = 39\ \mu m$ and $r = 32\ \mu m$, respectively, gap of circle split ring $S_c = 10\ \mu m$, width of silica (graphene) strip $w_2 = 10\ \mu m$, and the period of unit cell $P/2 = 90\ \mu m$; for all split rings, the thicknesses of the metallic layer, silica layer and substrate were $0.2\ \mu m$, $0.3\ \mu m$ and $1000\ \mu m$, respectively.

Figure 2 is the schematics of our device, which was formed of the unit cell array. We set aluminum electrodes in the edges of y axis on the front and back surfaces of the chip, which were linked to the pins of voltage control circuit Pin_A and Pin_B, respectively. The schematic of the whole device is shown in Figure 2a (note that actually the entire device has more than 50×50-unit cells in the xy-plane and is larger than the size of THz beam). Figure 2b,c strictly describe the structure and dimensions from the front and back views of the device, which are the period arrays of the unit cell shown in Figure 1b and 1c, respectively. As shown in Figure 2d, the metamaterials on the front and back sides were connected to the external voltage sources V_{CA} and V_{CB}, respectively, which could program the on–off commands independently. The separated control circuits could shift the Fermi levels of the graphene on the front and back sides individually, and therefore modulate the transmission amplitudes of dual-band determined by the double-sided metamaterials.

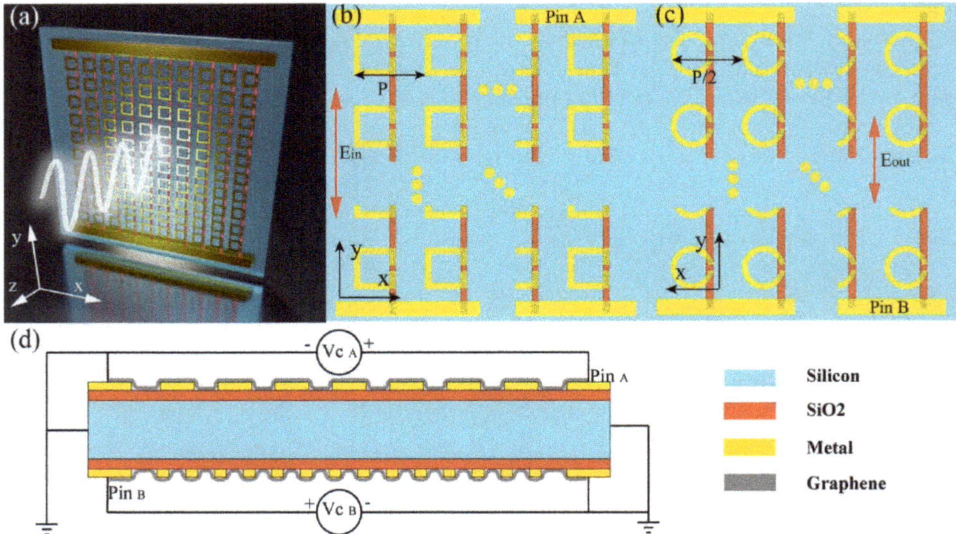

Figure 2. Structures of the device: (**a**) The schematic of the device; (**b**) the front view, (**c**) back view and (**d**) lateral view of the device.

We simulated the transmission spectra using CST microwave studio in time domain, and the incident THz waves were polarized along the y axis. The dielectric constants of silicon and silica are 11.69 and 3.46, respectively, and the conductivity of metallic rings made of aluminum is $3.6{\times}10^7$ S/m [25,26]. The conductivity of graphene can be expressed by the Drude-like model [27,28] $\sigma(\omega) = ie^2E_F/\left[\pi\hbar^2\left(\omega + i\tau^{-1}\right)\right]$, where e and $\hbar$ are electron charge and Planck constant, respectively, and E_F represents the Fermi level of graphene. The relaxation time τ can be calculated by $\tau = \mu E_F/\left(ev_F^2\right)$, and in the THz regime, we set the carrier mobility $\mu = 3000$ cm^2/V·s and the Fermi velocity $v_F = 1.1 \times 10^6$ m/s [27,28]. Hence, the conductivity of graphene can be varied by changing the Fermi level via controlling external voltage [19,29].

Considering the manufacturing process, the chip can be feasibly fabricated by means of conventional techniques. The fabrication of the chip is divided into four steps, roughly: Firstly, grow metallic structures on the front and back surfaces of the silicon/silicon dioxide (SSD) wafer using photolithography and lift-off process [3]; secondly, the photoresist patterns and the metallic structures together serve as a mask during the reactive ion etching (RIE) of the unneeded silica structures on both sides of the SSD wafer [30]; thirdly, transfer the graphene to both sides of the chip and then pattern the graphene strips using e-beam lithography and etching with a radio frequency oxygen plasma asher [19]; finally, connect the electrodes to the external voltage source through leads [19].

3. Results and Discussion

3.1. Equivalent Circuit for the Resonators

To better understand the physical mechanism affecting the performance of the encoder, we investigated the characteristics of the hybrid metamaterials on the front and back surfaces of the device separately. Figure 3a shows the normalized transmission spectra of the front side metamaterial. We simulated the results by varying the Fermi level of the graphene strips E_F from 0.01 eV to 1 eV. When the E_F was 0.01 eV, the resonance at 0.20 THz (band A) was obviously excited, however, as the E_F increased, the resonance gradually weakened, and the bandwidth broadened. When the E_F increased to 0.8 eV or higher, the resonance was almost vanished. Associated with the distributions of the surface current density in the metallic layer shown in Figure 3b,c, we can see that the loop current was evidently stimulated within the square split ring when $E_F = 0.01$ eV; as for the case when $E_F = 0.8$ eV, the distribution of current became symmetrical along the y axis. Similar results can be found in the transmission spectra and the distributions of the surface current density of the back-side metamaterial shown in Figure 3d–f, except that the resonance frequency was shifted to 0.33 THz (band B).

The behavior of the resonators can be explained with the theory of equivalent circuit [31,32]. For a metallic structure of split ring resonator (SRR), its gap capacitance and loop inductance can be described by an equivalent capacitance C_e and an equivalent inductance L_e, respectively. In our work, when the voltage (the difference between the gate voltage and the voltage shifting the Fermi level to the Dirac point [22]) is not applied, the Fermi level of graphene is on the Dirac point. Therefore, the conductivity of graphene is very low (i.e., the graphene is hardly electrically conductive) [33]. Hence, the gap in the square/circle split ring is non-conducting. In this situation, the loop current shown in Figure 3b/e for the square/circle split ring with $E_F = 0.01$ eV can be modeled with LC equivalent circuit, which is diagramed in Figure 3g/h, corresponding to the situation that the switch $k_{f/b}$ is open. Thus, the resonance frequency f of SRR can be expressed as [31,32]:

$$f = \frac{1}{2\pi \sqrt{L_e C_e}} \tag{1}$$

Here, the equivalent inductance L_e is approximately proportional to the size of the SRR [34]. Due to the different shapes of the resonator, the size can be reasonably converted with the area. Consequently, the resonance frequency can be qualitatively estimated with $f_s \sim 1/L$ for the square split ring and $f_c \sim 1/R$ for the circle split ring. Since R is smaller than L, the resonance frequency of the circle split ring will be higher than that of the square split ring, which agrees with our results shown in Figure 3a,d. Besides, the equivalent capacitance C_e will also be affected by other parameters (gap, width or thickness of the metallic structure) [34]. As we mentioned in Section 2, we optimized the geometrical parameters of two resonators in order to obtain the good performance in modulation depth.

When the voltage is applied on the graphene strip, the Fermi level is shifted away from the Dirac point [33]. Thus, the conductivity of graphene strip becomes large, and graphene switches to be electrically conductive. Hence, the gap in the square/circle split ring is conducting to a certain extent, and the strength of the resonance will generally weaken. When the Fermi level is tuned to 0.8 eV, the equivalent circuit for the square/circle split ring, as shown in Figure 3g/h, will be altered to the RLC parallel resonance circuit, corresponding to the case that the switch $k_{f/b}$ is closed. Here, $r_{f/b}$ is the equivalent resistance between the gap and $R_{f/b}$ is the equivalent resistance between the unit cells. Due to the high conductivity of graphene, the initial LC resonance is broken: each capacitator is connected in parallel with a low resistance, and the equivalent resistance between the unit cells is also reduced. As a result, the resonance behavior is changed to the symmetrical current distribution shown in Figure 3c/f, which corresponds to the dipole mode whose resonance frequency will be blue shifted. Note that if E_F further rises to 1 eV, the new merged dipole mode will be resonated (obviously around 0.30 THz in Figure 3a for the square split ring) due to the metal-like conductivity of the graphene. As a result, the transmission amplitude at higher frequencies will decrease, which will make trouble for designing dual-band devices. This is why we choose 0.8 eV as the higher Fermi level corresponding to "On" status for the encoder hereinafter.

For the cases that Fermi level is between 0.01 eV and 0.8 eV, the modulation of the transmission dip also can be qualitatively explained with the theory of equivalent circuit. Take square split ring as an example: According to the relation of quality factor Q to the parameters of the equivalent circuit $Q = r \sqrt{C_e/L_e}$, when the voltage is applied, the equivalent circuit is switched from initial LC resonance circuit to RLC parallel resonance, and the parallel resistance r_f varies from infinity (because of the open state of the switch k_f) to a finite value. Consequently, Q will be reduced and the resonance at band A will be weakened. These coincide with the simulated results shown in Figure 3a: When Fermi level increases, the transmission gradually enlarges, and the bandwidth gradually broadens. The similar analysis also can be applied to explain the results of the circle split ring and we will not discuss in detail.

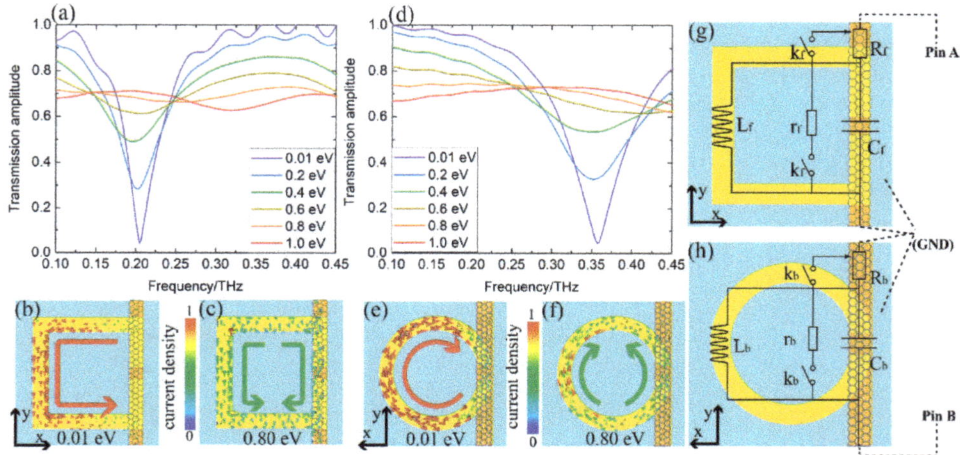

Figure 3. (a) Transmission spectra, distributions of the surface current density with (b) $E_F = 0.01$ eV and (c) $E_F = 0.8$ eV of the single metamaterial on the front surface. (d) Transmission spectra, distributions of the surface current density with (e) $E_F = 0.01$ eV and (f) $E_F = 0.8$ eV of the single metamaterial on the back surfaces. Equivalent circuit diagrams of the resonator in the (g) front and (h) back-side metamaterials.

3.2. Performance of the Encoder

Now, if we choose $E_F = 0.01$ eV as the low transmission state, and $E_F = 0.8$ eV as the high transmission state, which corresponding to the situations without and with applying the voltage on the graphene stripes, respectively, we can obtain two different states for each resonator. In this way, we can realize one-bit coding at one band. To double the number of bits, we should assemble two resonators working at two different bands and meanwhile eliminate their coupling. We put the square and circle split ring arrays on the front and back surfaces of the metamaterials, respectively, so that the coupling between the two kinds of resonator can be effectively reduced. We have investigated the coupling effect by changing the thickness of the substrate. When the front and back side metamaterials were separated more than 500 μm, the coupling effect nearly disappeared. The reason for choosing 1 mm-thick substrate in our device is that when we detect the transmission spectrum using THz time domain system, we can get longer time signal and obtain more detailed information in the frequency domain after Fourier transform.

After determining the optimized geometrical parameters and Fermi level values, we simulated the performance of the encoder with double-sided hybrid metamaterials. Figure 4 demonstrates the simulated normalized transmission spectra of the device showing four encoding statuses in frequency domain, which corresponds to the different cases of the voltage control listed in Table 1. Here, for the status of the controller, "On" or "Off" refers to the situations that Pin_A/B) is loaded with or without voltages, respectively, and thus the corresponding Fermi level of the graphene is modeled as 0.01 eV or 0.8 eV, respectively; for the transmission amplitude, we define the value as "Low" or "High" depending on whether it is lower or higher than the threshold (chosen as 0.4). Apparently, we achieve four encoding statuses—00, 01, 10, 11—via modulating the transmission amplitude of bands A and B (resonated around 0.20 THz and 0.33 THz, respectively) by applying the voltages to the structures on the front and back surfaces independently. Comparing to the transmission spectra of each side metamaterials shown in Figure 3a,d, we can identify that the working bands A and B are determined by the metamaterials on the front and back surfaces, respectively. Since the frequencies of the resonance dips were barely moved, the coupling between the two metamaterials can be ignored. In order to measure the effectiveness of our device, we extracted the modulation depth from the transmission spectra (the value of modulation depth was calculated using the amplitude of codes 00 and 11). As denoted in Figure 4, the modulation depth could reach as high as 79.6%, which indicates our encoder has distinguished performance.

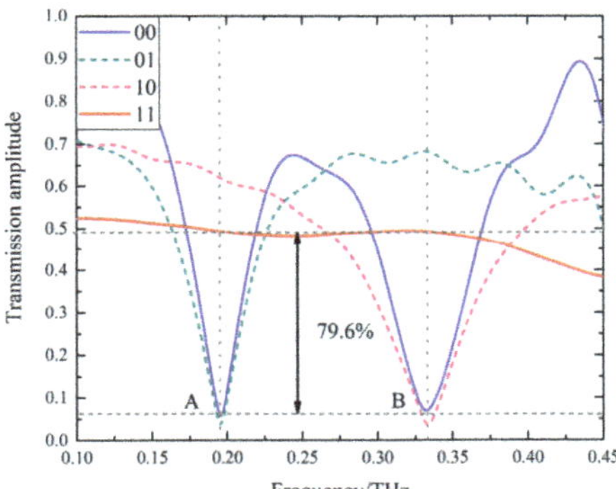

Figure 4. Normalized transmission of the device in different encoding cases.

Table 1. Encoding status and the corresponding status of the controller and the transmission.

Code	Voltage of Controller		Transmission Amplitude	
	Pin_A	**Pin_B**	**Band A**	**Band B**
00	Off [1]	Off	Low	Low
01	Off	On	Low	High
10	On [2]	Off	High	Low
11	On	On	High	High

[1] Status "Off" was simulated with $E_F = 0.01$ eV. [2] Status "On" was simulated with $E_F = 0.8$ eV.

Subsequently, we investigated the resonance details about different coding states. Figure 5a–d shows the distributions of the surface current density in codes 00, 01, 10 and 11, respectively, in which the top and bottom rows correspond to the resonances at 0.20 THz (band A) and 0.33 THz (band B), respectively. Visibly, band A was dominated by the resonance in the square split rings, and band B was governed by the resonance in the circle split rings. The resonance behaviors are almost identical to those shown in Figure 3b–f. It is obvious that the resonances are strong when there is no voltage applied to the structures (for example the situation of code 00), which coincide with the deep resonance dips in the transmission spectra. If the voltage is applied to the structures (for example the situation of code 11), the resonances are much weaker, and the resonance dips will disappear in the transmission spectra.

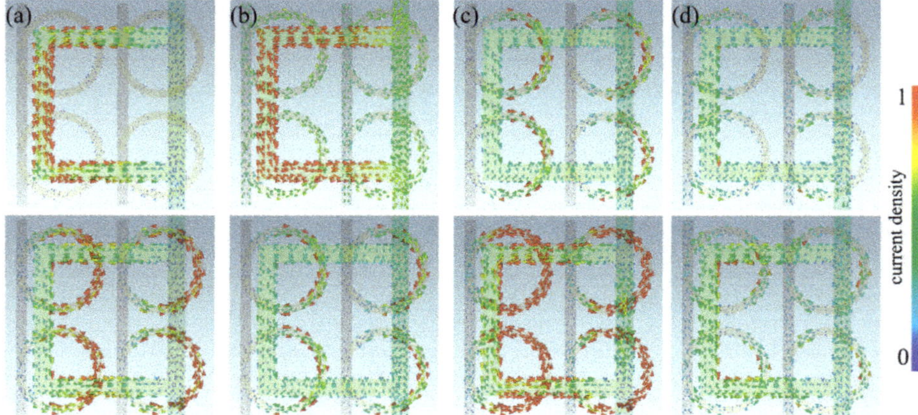

Figure 5. Distributions of surface current density in different codes at 0.20 THz (top row) and 0.33 THz (bottom row): (**a**) code 00, (**b**) code 01, (**c**) code 10, (**d**) code 11.

3.3. Further Discussion

At last, we briefly discuss the issues about the potential applications. The applied voltage may be a considerable indicator for graphene-based device. According to the references [16] and [22], we can roughly estimate that the required voltage shifting Fermi level to 0.8 eV is about three hundred volt. There are two possible methods to reduce the applied voltage. One is to choose lower Fermi level as the "On" state, and the price is that the modulation depth will be diminished. The other is to decrease the thickness of silica layer, and the required voltage for the same Fermi level may be effectively reduced, but the breakdown voltage of the device will also drop off. Those solutions need more investigations based on theory and experiment, which will be the key points in our future work.

According to our simulated results and discussion, we can infer that if the metamaterials laying on different layers are distant enough, the interaction between them can be almost ignored. Thus, if we arrange more resonators by means of multiplying the layers, the functionality of the device may be efficiently expanded. Our scheme of encoder has potential in developing multibit device combined with CMOS (Complementary Metal Oxide Semiconductor) technique in the future.

4. Conclusions

We propose a two-bit encoding device working at terahertz frequencies. The device assembles the two graphene-based hybrid metamaterials with different shapes deposited on the front and back surfaces of a silicon wafer. Using the external electrodes to change the conductivities of the graphene on each side individually, we realize the active control of dual-band transmission amplitudes. The results demonstrate that our device can function at 0.20 THz and 0.33 THz with good performance, and the modulation depth can reach 79.6%. Comparing with the present THz modulators, our encoder can not only maximally enhance the coding number in a single chip, but also guarantee the modulation efficiency. The functionality of the device may also be expanded by arranging more separately controlled resonators in multilayer structures. Moreover, the device proposed in this paper can also be widely used in other bands. Our encoder may pave a new way to develop multifunctional and integrated devices, such as multiband filter and sensor, frequency division multiplexer and logical circuitry, which will promote the development of THz communications and other applications.

5. Patents

Shan Yin, Xintong Shi, Wei Huang, Ling Guo, Fangrong Hu, Wentao Zhang, Xianming Xiong. A transmission-type terahertz two-bit encoding device and system. 2019210801324 (China).

Author Contributions: Conceptualization, S.Y.; investigation and data curation, X.S.; writing—original draft preparation, S.Y., X.S. and W.H.; writing—review and editing, W.H. and W.Z.; funding acquisition, S.Y., W.H., W.Z., F.H. and Z.Q.; project administration, X.X.

Funding: This research was funded by the National Science and Technology Major Project of China (No. 2017ZX02101007-003); the National Natural Science Foundation of China (Nos. 61565004, 61665001 and 61965005); the Natural Science Foundation of Guangxi Province (Nos. 2017GXNSFBA198116 and 2018GXNSFAA281163); the Science and Technology Program of Guangxi Province (No. 2018AD19058); and the Open Project of Guangxi Colleges and Universities Key Laboratory of Complex System Optimization and Big Data Processing (No. 2017CSOBDP0203). Wei Huang acknowledges the funding from Guangxi Oversea 100 Talent Project and Wentao Zhang acknowledges the funding from Guangxi Distinguished Expert Project.

Conflicts of Interest: The authors declare no conflict of interest.

References

1. Ferguson, B.; Zhang, X.C. Materials for terahertz science and technology. *Nat. Mater.* **2002**, *1*, 26. [CrossRef] [PubMed]
2. Dhillon, S.S.; Vitiello, M.S.; Linfield, E.H.; Davies, A.G.; Hoffmann, M.C.; Booske, J.; Paoloni, C.; Gensch, M.; Weightman, P.; Williams, G.P.; et al. The 2017 terahertz science and technology roadmap. *J. Phys. D Appl. Phys.* **2017**, *50*, 043001. [CrossRef]
3. Chen, H.-T.; O'Hara, J.F.; Azad, A.K.; Taylor, A.J.; Averitt, R.D.; Shrekenhamer, D.B.; Padilla, W.J. Experimental demonstration of frequency-agile terahertz metamaterials. *Nat. Photonics* **2008**, *2*, 295–298. [CrossRef]
4. Yu, N.; Capasso, F. Flat optics with designer metasurfaces. *Nat. Mater.* **2014**, *13*, 139–150. [CrossRef] [PubMed]
5. Li, C.; Chang, C.C.; Zhou, Q.; Zhang, C.; Chen, H.T. Resonance coupling and polarization conversion in terahertz metasurfaces with twisted split-ring resonator pairs. *Opt. Express* **2017**, *25*, 25842–25852. [CrossRef] [PubMed]
6. Yin, S.; Hu, F.; Chen, X.; Han, J.; Chen, T.; Xiong, X.; Zhang, W. Ruler equation for precisely tailoring the resonance frequency of terahertz U-shaped metamaterials. *J. Opt.* **2019**, *21*, 025101. [CrossRef]
7. Hu, F.; Guo, E.; Xu, X.; Li, P.; Xu, X.; Yin, S.; Wang, Y.; Chen, T.; Yin, X.; Zhang, W. Real-timely monitoring the interaction between bovine serum albumin and drugs in aqueous with terahertz metamaterial biosensor. *Opt. Commun.* **2017**, *388*, 62–67. [CrossRef]
8. Huang, W.; Rangelov, A.A.; Kyoseva, E. Complete achromatic optical switching between two waveguides with a sign flip of the phase mismatch. *Phys. Rev. A* **2014**, *90*, 053837. [CrossRef]
9. Chen, X.; Ghosh, S.; Xu, Q.; Ouyang, C.; Li, Y.; Zhang, X.; Tian, Z.; Gu, J.; Liu, L.; Azad, A.K.; et al. Active control of polarization-dependent near-field coupling in hybrid metasurfaces. *Appl. Phys. Lett.* **2018**, *113*, 061111. [CrossRef]
10. Chen, H.-T.; Padilla, W.J.; Cich, M.J.; Azad, A.K.; Averitt, R.D.; Taylor, A.J. A metamaterial solid-state terahertz phase modulator. *Nat. Photonics* **2009**, *3*, 148–151. [CrossRef]
11. Li, H.; Cao, Q.; Liu, L.; Wang, Y. An Improved Multifunctional Active Frequency Selective Surface. *IEEE Trans. Antennas Propag.* **2018**, *66*, 1854–1862. [CrossRef]
12. Wang, L.; Zhang, Y.; Guo, X.; Chen, T.; Liang, H.; Hao, X.; Hou, X.; Kou, W.; Zhao, Y.; Zhou, T.; et al. A Review of THz Modulators with Dynamic Tunable Metasurfaces. *Nanomaterials* **2019**, *9*, 965. [CrossRef] [PubMed]
13. Jnawali, G.; Rao, Y.; Yan, H.; Heinz, T.F. Observation of a transient decrease in terahertz conductivity of single-layer graphene induced by ultrafast optical excitation. *Nano Lett.* **2013**, *13*, 524–530. [CrossRef] [PubMed]
14. Yin, Y.; Cheng, Z.; Wang, L.; Jin, K.; Wang, W. Graphene, a material for high temperature devices–intrinsic carrier density, carrier drift velocity, and lattice energy. *Sci. Rep.* **2014**, *4*, 5758. [CrossRef] [PubMed]
15. Huang, W.; Liang, S.-J.; Kyoseva, E.; Ang, L.K. Adiabatic control of surface plasmon-polaritons in a 3-layers graphene curved configuration. *Carbon* **2018**, *127*, 187–192. [CrossRef]
16. Dabidian, N.; Kholmanov, I.; Khanikaev, A.B.; Tatar, K.; Trendafilov, S.; Mousavi, S.H.; Magnuson, C.; Ruoff, R.S.; Shvets, G. Electrical Switching of Infrared Light Using Graphene Integration with Plasmonic Fano Resonant Metasurfaces. *ACS Photonics* **2015**, *2*, 216–227. [CrossRef]
17. Farmer, D.B.; Avouris, P.; Li, Y.; Heinz, T.F.; Han, S.J. Ultrasensitive Plasmonic Detection of Molecules with Graphene. *ACS Photonics* **2016**, *3*, 553–557. [CrossRef]

18. Li, Q.; Tian, Z.; Zhang, X.; Singh, R.; Du, L.; Gu, J.; Han, J.; Zhang, W. Active graphene-silicon hybrid diode for terahertz waves. *Nat. Commun.* **2015**, *6*, 7082. [CrossRef]

19. Kindness, S.J.; Almond, N.W.; Wei, B.; Wallis, R.; Michailow, W.; Kamboj, V.S.; Braeuninger-Weimer, P.; Hofmann, S.; Beere, H.E.; Ritchie, D.A.; et al. Active Control of Electromagnetically Induced Transparency in a Terahertz Metamaterial Array with Graphene for Continuous Resonance Frequency Tuning. *Adv. Opt. Mater.* **2018**, *6*, 1800570. [CrossRef]

20. Kleine-Ostmann, T.; Nagatsuma, T. A Review on Terahertz Communications Research. *J. Infrared Millim. Terahertz Waves* **2011**, *32*, 143–171. [CrossRef]

21. Naeem, N.; Parveen, S.; Ismail, A. Terahertz Communications for 5G and Beyond. In *Antenna Fundamentals for Legacy Mobile Applications and Beyond*; Elfergani, I., Hussaini, A.S., Rodriguez, J., Abd-Alhameed, R., Eds.; Springer International Publishing: Cham, Switzerland, 2018; pp. 305–322.

22. Gao, W.; Shu, J.; Reichel, K.; Nickel, D.V.; He, X.; Shi, G.; Vajtai, R.; Ajayan, P.M.; Kono, J.; Mittleman, D.M.; et al. High-Contrast Terahertz Wave Modulation by Gated Graphene Enhanced by Extraordinary Transmission through Ring Apertures. *Nano Lett.* **2014**, *14*, 1242–1248. [CrossRef] [PubMed]

23. Fan, Y.; Qian, Y.; Yin, S.; Li, D.; Jiang, M.; Lin, X.; Hu, F. Multi-band tunable terahertz bandpass filter based on vanadium dioxide hybrid metamaterial. *Mater. Res. Express* **2019**, *6*, 055809. [CrossRef]

24. Xiao, S.; Wang, T.; Liu, T.; Yan, X.; Li, Z.; Xu, C. Active modulation of electromagnetically induced transparency analogue in terahertz hybrid metal-graphene metamaterials. *Carbon* **2018**, *126*, 271–278. [CrossRef]

25. Srivastava, Y.K.; Manjappa, M.; Cong, L.; Cao, W.; Al-Naib, I.; Zhang, W.; Singh, R. Ultrahigh-Q Fano Resonances in Terahertz Metasurfaces: Strong Influence of Metallic Conductivity at Extremely Low Asymmetry. *Adv. Opt. Mater.* **2015**, *4*, 457–463. [CrossRef]

26. Ordal, M.A.; Long, L.L.; Bell, R.J.; Bell, S.E.; Bell, R.R.; Alexander, R.W.; Ward, C.A. Optical properties of the metals Al, Co, Cu, Au, Fe, Pb, Ni, Pd, Pt, Ag, Ti, and W in the infrared and far infrared. *Appl. Opt.* **1983**, *22*, 1099–1120. [CrossRef]

27. Liu, T.; Zhou, C.; Cheng, L.; Jiang, X.; Wang, G.; Xu, C.; Xiao, S. Actively tunable slow light in a terahertz hybrid metal-graphene metamaterial. *J. Opt.* **2019**, *21*, 035101. [CrossRef]

28. Wang, X.; Meng, H.; Deng, S.; Lao, C.; Wei, Z.; Wang, F.; Tan, C.; Huang, X. Hybrid Metal Graphene-Based Tunable Plasmon-Induced Transparency in Terahertz Metasurface. *Nanomaterials* **2019**, *9*, 385. [CrossRef]

29. Gomez-Diaz, J.S.; Moldovan, C.; Capdevila, S.; Romeu, J.; Bernard, L.S.; Magrez, A.; Ionescu, A.M.; Perruisseau-Carrier, J. Self-biased reconfigurable graphene stacks for terahertz plasmonics. *Nat. Commun.* **2015**, *6*, 6334. [CrossRef]

30. Gaboriau, F.; Cartry, G.; Peignon, M.C.; Cardinaud, C. Selective and deep plasma etching of SiO_2: Comparison between different fluorocarbon gases (CF_4, C_2F_6, CHF_3) mixed with CH_4 or H_2 and influence of the residence time. *J. Vac. Sci. Technol. B Microelectron. Nanometer Struct.* **2002**, *20*, 1514. [CrossRef]

31. Shamonin, M.; Shamonina, E.; Kalinin, V.; Solymar, L. Properties of a metamaterial element: Analytical solutions and numerical simulations for a singly split double ring. *J. Appl. Phys.* **2004**, *95*, 3778–3784. [CrossRef]

32. Shelton, D.J.; Brener, I.; Ginn, J.C.; Sinclair, M.B.; Peters, D.W.; Coffey, K.R.; Boreman, G.D. Strong coupling between nanoscale metamaterials and phonons. *Nano Lett.* **2011**, *11*, 2104–2108. [CrossRef] [PubMed]

33. Novoselov, K.S.; Geim, A.K.; Morozov, S.V.; Jiang, D.; Zhang, Y.; Dubonos, S.V.; Grigorieva, I.V.; Firsov, A.A. Electric Field Effect in Atomically Thin Carbon Films. *Science* **2004**, *306*, 666–669. [CrossRef] [PubMed]

34. Qun, W.; Ming-Feng, W.; Fan-Yi, M.; Jian, W.; Le-Wei, L. Modeling the effects of an individual SRR by equivalent circuit method. In Proceedings of the 2005 IEEE Antennas and Propagation Society International Symposium, Washington, DC, USA, 3–8 July 2005; pp. 631–634.

Article

Microfluidic Sensor Based on Composite Left-Right Handed Transmission Line

Vasa Radonić *, Slobodan Birgermajer, Ivana Podunavac, Mila Djisalov, Ivana Gadjanski and Goran Kitić

BioSense Institute - Research and development institute for information technologies in biosystems, University of Novi Sad, 21101 Novi Sad, Serbia; b.sloba@biosense.rs (S.B.); ivana.podunavac@biosense.rs (I.P.); mila.mandic@biosense.rs (M.D.); igadjanski@biosense.rs (I.G.); gkitic@biosense.rs (G.K.)
* Correspondence: vasarad@biosense.rs; Tel.: +381-63-518-335

Received: 31 October 2019; Accepted: 1 December 2019; Published: 4 December 2019

Abstract: In this paper, we propose a novel metamaterial-based microfluidic sensor that permits the monitoring of properties of the fluid flowing in the microfluidic reservoir embedded between the composite left–right handed (CLRH) microstrip line and the ground plane. The sensor's working principle is based on the phase shift measurement of the two signals, the referent one that is guided through conventional microstrip line and measurement signal guided through the CLRH line. At the operating frequency of 1.275 GHz, the CLRH line supports electromagnetic waves with group and phase velocities that are antiparallel, and therefore the phase "advance" occurs in the case of CLRH line, while phase delay arises in the right-handed (RH) frequency band. The change of the fluid's properties that flow in the microfluidic reservoir causes the change of effective permittivity of the microstrip substrate, and subsequently the phase velocity changes, as well as the phase shift. This effect was used in the design of the microfluidic sensor for the measurement of characteristics of the fluid that flows in the microfluidic reservoir placed under the CLRH line. The complete measurement system was developed including the Wilkinson power divider that splits the signal between conventional RH and CLRH section, transmission lines with the microfluidic reservoirs, and a detection circuit for phase shift measurement. Measurement results for different fluids confirm that the proposed sensor is characterized by relatively high sensitivity and good linearity (R^2 = 0.94). In this study, the practical application of the proposed sensor is demonstrated for the biomass estimation inside the microfluidic bioreactors, which are used for the cultivation of MRC-5 fibroblasts.

Keywords: metamaterials; left-handed line; sensors; phase shift

---

1. Introduction

In the last two decades, innovative results have been achieved in the field of metamaterials, namely, artificial structures that exhibit electromagnetic, acoustical, and optical properties that are generally not found in nature. From the moment of the first experimental verification of the single negative metamaterials [1,2], the unique properties based on negative permittivity, permeability or index of refraction have found a place in a number of novel devices and applications [3–8]. Special attention has been given to double-negative or left-handed (LH) media, which at the same time show negative values of permittivity and permeability in a certain frequency range.

Considerable attention has been focused on the implementation of metamaterials in sensor designs for material characterization applications, as well as on their integration with microfluidic devices [9,10]. Although a number of sensor solutions have been proposed so far, all proposed designs operate according to the resonant effect that relays resonant frequency changes. In this paper, we propose a metamaterials-based sensor that utilizes a transmission line (TL) concept and operates at one specific frequency. The special case is concerning 1D metamaterials in which the fundamental electromagnetic properties of the left-handed media can be modelled using a conventional transmission line (TL) theory [11]. The TL approach, which is based on the composite left–right handed (CLRH) line, provides good insight into physical phenomena. The general CLRH model consists of an ideal LH section that can be modelled as the combination of a per-unit length series capacitance and a per-unit length shunt inductance, and a parasitic RH section (i.e., a parasitic per-unit length series inductance and a per-unit length shunt capacitance). The CLRH line has two propagation bands, one left-handed and one right-handed. The group velocity and phase velocity of CLRH calculated from the dispersion diagram in [11] show that group and phase velocity are parallel in the RH band and antiparallel in the case of the LH one. Therefore, phase advance happens in the LH frequency range, while phase delay arises in the RH frequency range. The first experimental verification of the 1D concept of CLRH is demonstrated using microstrip line configuration in [11]. The phase response in the case of the microstrip CLRH is nonlinearly dependent on the propagation constant in the LH band, and it is strongly influenced by substrate properties (i.e., its permittivity and permeability). This concept was utilized in the design of the novel microstrip microfluidic sensor that operates according to the phase shift measurement.

In this paper, we propose a novel metamaterial-based microfluidic sensor that permits the monitoring of the fluid properties inside the microfluidic reservoir embedded between the CLRH TL line and the ground plane. In the case of the microstrip configuration, the properties of the dielectric substrate have a strong influence on the phase response. The operating principle is based on the transmission phase shift measurement of the two signals, the referent one guided through the conventional microstrip RH line section and the signal guided through the CLRH line. The TL method is one of the commonly used methods appropriate for material characterization in a wider frequency band. Therefore, a number of TL configurations have been proposed for the characterization of different hard solid, liquid or powder materials [12–19]. The TL method comprises measurements of both reflection and transmission characteristics and their combination, and can be used for characterization of permittivity as well as permeability. One of the relatively fast and simple TL methods for the determination of dielectric properties of material is a method based on the measurement of the phase shift of the transmitted signal. Compared with the other TL methods, phase shift methods are less sensitive to noise [19,20], which permits characterization at a single frequency and simplifies the development of supporting electronics and easy integration with the sensor element, as well as allowing the fabrication of low-cost in-field sensing devices. For that reason, TL methods have found application in different sensor designs such as soil moisture sensors [21], microfluidic sensors for detection of fluid mixture efficiency [22], and so on.

The proposed method based on phase shifting will be explained in detail using our CLRH microstrip microfluidic sensor configuration as an example. The permittivity of the fluid inside the microfluidic channel will be correlated by measuring the phase shift of the signal that passes thought the CLRH sensing section. A detailed description is provided. of the complete measurement system, including the design of the microstrip Wilkinson power divider, which splits the signal between the conventional-referent section and the CLRH section, microfluidic reservoirs, and a detection circuit for the measurement of the phase difference. The proposed microfluidic sensor has been manufactured using hybrid microfluidic fabrication technology that combines a laser micromachining process, the xurography technique, and lamination, [23]. The detection electronic circuit has been developed using printed circuit board (PCB) technology. The sensor's performances have been evaluated based on the measurement of different fluids inside the microfluidic channel. To demonstrate applicability, the proposed sensor has been used to evaluate the biomass of the MRC-5 fibroblast cells grown in the microfluidic bioreactor.

2. Materials and Methods

2.1. Materials

The proposed microfluidic chip was manufactured using an 80-µm thick polyvinyl chloride (PVC) lamination foil (MBL 80MIC Belgrade, Serbia) and a 2-mm thick poly(methyl methacrylate) (PMMA). Double-sided 3M 9088 tape was used for microfluidic chip bounding. Conductive aluminum foil 3M 3302 was used for the realization of the microchip TL lines and the sensor ground plane. RS PRO silver conductive adhesive epoxy was used for mounting vias, resistor, and surface mount assembly (SMA) connectors.

Deionized water (Grade 2 by ISO 3696 (1987)), isopropyl (99.7%, Sigma-Aldrich, Saint Louis, MO, USA), methanol (99.8%, Sigma-Aldrich), ethanol (99.8%, Sigma-Aldrich), and sunflower oil (Vital, Vrbas, Serbia) were used as fluids in the microfluidic channel.

The cell line used in this study for cultivation in the microfluidic bioreactor was MRC-5 (human fibroblasts, American Type Culture Collection CCL 171). The cells were grown in Dulbecco's modified Eagle's medium (DMEM) with 4.5% of glucose, supplemented with 10% of fetal calf serum (Sigma-Aldrich) and antibiotic/antimycotic solution (Sigma-Aldrich). The single cell suspension was obtained using 0.25% trypsin in EDTA (Serva). Cells were harvested and counted by 0.1% trypan blue exclusion. The viability of cells used in the assay was over 90%.

The phase detector module for phase shift difference measurement was realized using integrated circuit AD8302 Analogue Devices on the PCB board with supporting electronic components according to the manufacturer's recommendation [24].

2.2. Equipment and Small Tools

PMMA layer was cut with a CO_2 Gravograph LS1000XP laser. A plotter cutter (CE6000-60 PLUS, Graphtec America, Inc., Irvine, CA, USA) with a 45 cutting blade (CB09U) and cutting mat (12″ Silhouette Cameo Cutting Mat, Sacramento, USA) was used for carving inlets, outlets, and edges of PVC layers for microfluidic chips. Bondage between PVC and PMMA was performed through lamination with the uniaxial press (Carver 3895CEB, Wabash, USA). Aluminum tapes were cut with a Rofin-Sinar Power Line D-100 laser (Germany).

A vector network analyzer (VNA) E5071C Agilent Technology was used to measure the complex S-parameters (scattering parameters) and generate an input signal for phase shift measurement. An MSP430FR6989 LaunchPad Development Kit (Texas Instruments, USA) was used for the measurement of the output voltage.

2.3. Sensor Fabrication

The proposed microfluidic chip was composed of five dielectric layers. The top and bottom layers were manufactured using PVC foils, the middle one was realized in PMMA, and the two bonding layers between them were made using double-sided adhesive tape; two conductive layers (a microstrip line on the top and a ground plane from the bottom side) were realized using aluminum sticky foil, as shown in Figure 1. In the first step, the CO_2 laser was used for cutting the middle chip layer with the microfluidic reservoir in the PMMA using a power of 60 W and a feed speed of 20 mm/s. Top and bottom PVC covers with the inlet and outlet of the microfluidic channels were cut using plotter cutter, as is the standard xurography technique. The cutting speed was set to 30 cm/s, while cutting force was set to 19 (range 1–38).

A multilayered chip was created by bonding PVC layers from the top and bottom side on PMMA using double-sided 3D adhesive tape in the uniaxial press with a pressure of 400 kg and a temperature of 70 °C for 1 min. Next, the conductive aluminum tapes were cut with the Nd:YAG laser with a current of 27.6 mA, a frequency of 10 kHz, and a cutting speed of 30 mm/s, and assembled on the top and bottom side of the microfluidic chip. Finally, the vias that connected inductive stubs and the ground plane, SMA connectors, and 100 Ω resistor for the Wilkinson power divider, were mounted using RS PRO silver conductive adhesive epoxy.

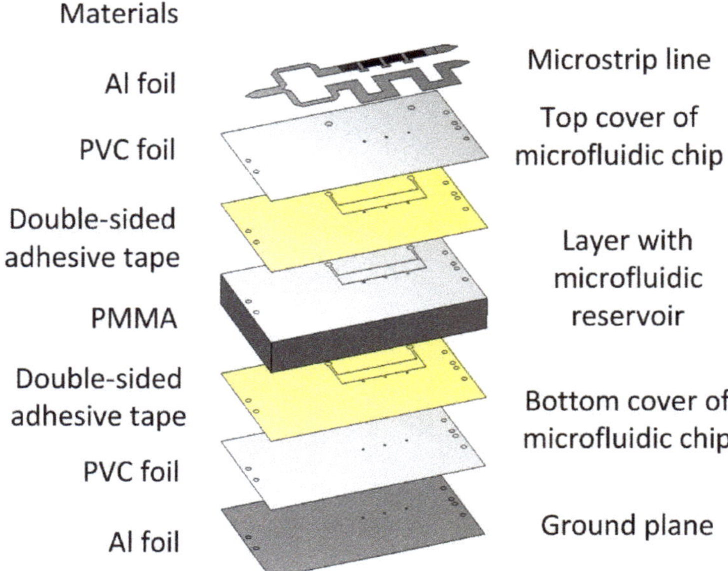

Figure 1. Layout of the proposed metamaterial-based microfluidic sensor with all relevant layers and materials (PVC- polyvinyl chloride, PMMA- poly(methyl methacrylate), Al-aluminum).

3. Configuration of the Proposed Sensor

The block diagram of the proposed sensor with supporting electronics is shown in Figure 2. It consists of a Wilkinson power divider that splits the input signal from the microwave oscillator between two TLs (the conventional RH TL and the CLRH TL), and a phase detector that converts the phase shift into the output voltage. The sinusoidal signal generated by the VNA is divided by the Wilkinson divider into measurement and referent signals. The measurement signal propagates along the CLRH TL section where the different phase shifts can be detected at the output for different fluids inside the microfluidic channel as a result of changes in the effective permittivity of the medium. The referent signal from the output of the RH TL and the signal from the CLRH sensor are fed into the input of the phase detector, which compares their phase responses. The voltage on the output of the phase detector is proportional to the phase shift of the input signals and is in relation to a dielectric constant in the microfluidic reservoir, as will be discussed below.

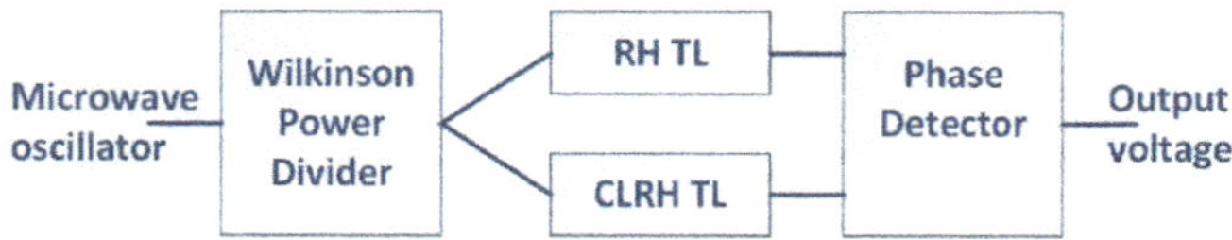

Figure 2. Block diagram the proposed microfluidic CLRH sensor (RH LT–right handed transmission line, CLRH TL–composite left-right handed transmission line).

The propagation signal propagates thought two sections: the RH TL and the CLRH TL, shown in Figure 3 together Wilkinson with a power divider. The equivalent circuit model of the conventional RH and CLRH microstrip TLs is shown in Figure 4a,b, respectively [11]. RH line was modelled using per-unit length series inductance ($L_R{'}$) and per-unit length shunt capacitance ($C_R{'}$), while the CLRH circuit model consists of a purely LH section in the form of per-unit length series capacitance ($C_{LH}{'}$), as well as a per-unit length shunt inductance ($L_{LH}{'}$) and parasitic RH section. The phase shift of the signal propagating through both TLs can be expressed as

$$\Delta\varphi = \frac{\omega l}{v_p},$$ (1)

and it is determined by phase velocity (v_p), angular frequency of the propagating signal (ω), and the physical properties of the transmission line (l). The phase velocity depends on the dispersion (β):

$$v_p = \frac{\omega}{\beta}.$$ (2)

For the RH TL, the dispersion relation is defined as

$$\beta_R = \frac{1}{\sqrt{L'_R C'_R}}$$ (3)

while for the CLRH it can be expressed as

$$\beta_{CLRH} = s(\omega)\sqrt{\omega^2 L'_{RH} C'_{RH} + \frac{1}{\omega^2 L'_{LH} C'_{LH}} - \left(\frac{L'_{RH}}{L'_{LH}} + \frac{C'_{RH}}{C'_{LH}}\right)}$$ (4)

where

$$s(\omega) = \begin{cases} -1 & if\ \omega < \omega_{\Gamma 1} = min\left(\frac{1}{\sqrt{L'_R C'_L}}, \frac{1}{\sqrt{L'_L C'_R}}\right) \\ 1 & if\ \omega > \omega_{\Gamma 2} = max\left(\frac{1}{\sqrt{L'_R C'_L}}, \frac{1}{\sqrt{L'_L C'_R}}\right) \end{cases}$$ (5)

From the dispersion diagram shown in Figure 4c for both transmission lines, it can be seen that in the case of the RH TL, the propagation constant is real and positive. On the other hand, for the CLRH TL the propagation constant can be purely real or purely imaginary. If the β_{CLRH} is purely real then propagation occurs, while stopband will exist for the imaginary value. Therefore, the CLRH line has two propagation bands in the lower LH and the upper RH. By combining Equation (3) into (1) we can see that phase shift is a linear function of the physical length in the case of the RH TL. Alternately, in the case of the CLRH TL, the phase shift changes intensely in the LH band and therefore has a stronger influence on the phase shift. In the used topology, the impact of the changes in fluid permittivity in the microfluidic channel is reflected in total capacitance of the CLRH TL and, consequently, will influence the phase velocity and the phase shift.

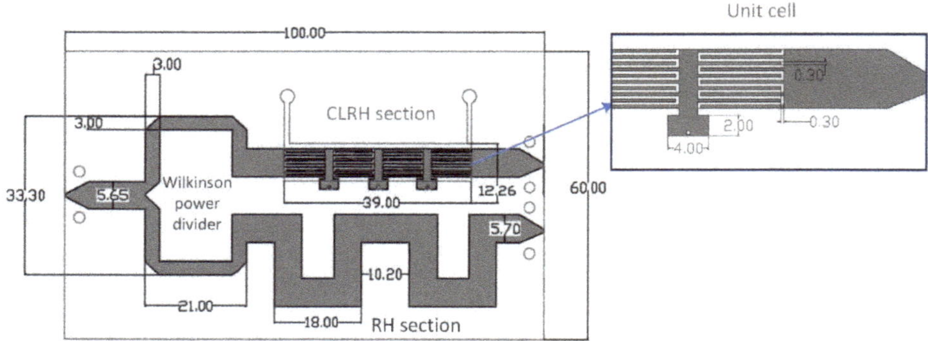

Figure 3. Top conductive layer with optimized dimensions in millimeters.

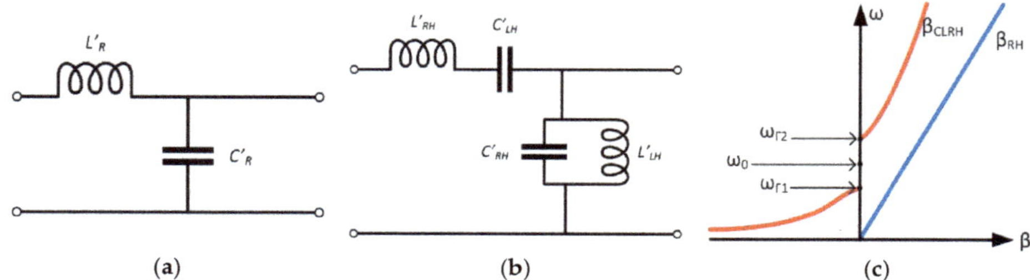

Figure 4. (**a**) Equivalent circuit of the RH TL. (**b**) Equivalent circuit of the CLRH TL. (**c**) Dispersion diagram of the RH and CLRH TLs.

The proposed metamaterials-based microfluidic sensor was manufactured as a combination of the proposed sensing concept and microfluidic platform with a reservoir placed under the CLRH section. The 3D view of the proposed microstrip microfluidic sensor is shown in Figure 1, while Figure 3 shows the optimized dimensions of transmission lines and the Wilkinson power divider, which is integrated together with two TL sections. The complete system is designed to operate at a frequency of 1.275 GHz. This frequency is high enough to neglect the effect of conductance on the phase shift, but also low enough to allow the realization of the electronic read-out using standard electronics components and commercially available integrated circuits. The RH microstrip line section was made as a meandered line, while the CLRH section was designed using three-unit cells comprised of per-unit length serial capacitance in the form of the interdigital capacitor, and per-unit length shunt inductance in the form of shunted inductive stub. Dimensions of the unit cell are optimized to provide LH behavior at a frequency of 1.275 GHz. In contrast to the standard phase comparator, a conventional microstrip line was replaced with the CLRH transmission line. In that manner, the phase "advance" occurs in the case of the CLRH line, while phase delay occurs in the RH frequency band, due to the backward wave propagation in the LH range [11]. Due to such an improved phase shift, the response can be obtained with better sensitivity at the output of the sensor, as will be discussed in more detail below.

The Wilkinson power divider and RH and LH TLs were designed separately using Sonnet software [25] and CST Microwave studio [26]. The optimized dimensions of these components are shown in Figure 3. In all simulations, the PVC foil was modelled with a permittivity of 3.1 and dielectric losses of 0.1, while the PMMA was modelled with permittivity of 3 and dielectric losses of 0.02 at a frequency of 1 GHz.

The response of the designed power divider is shown in Figure 5. It can be seen that, at an operating frequency of 1.275 GHz, the insertion losses of the transmitted signals for the proposed configuration are about −3.5 dB, while the isolation between two paths is higher than 40 dB.

The amplitude and phase characteristics of the referent RH line and CLRH line above the reservoir with various fluids in the reservoir are shown in Figure 6. The propagation constant and phase characteristics of the CLRH depend on the characteristics of the microstrip substrate (i.e., its dimensions and dielectric constant). Therefore, the change of the fluid in the microfluidic reservoir will influence the effective permittivity of the substrate and consequently the phase response [27]. For that reason, the central resonance of the LH band slightly shifts (Figure 6a), while the slope of the phase characteristics changes intensely (Figure 6b) when the properties of the fluid that flows through the microfluidic channel change. The phase shown in Figure 6b presents an unwrapped phase of the transmitted signals. The simulations have been performed for different fluids in the microfluidic reservoir, where the fluids are represented using their dielectric permittivity (ε_r) and dielectric loss tangent ($tg\delta$) (whose values are also given in the Figure 6a). The amplitude difference between two signals (i.e., the referent and measurement signals) is lower than 10 dB in the worst case at the operating frequency. Based on the above, we conclude that the commercially available phase detector can be used for phase shift measurement. In this study, an AD8302 phase comparator [24] was used to measure the phase shift. The RH line designed on PVC/PMMA substrate combination was optimized to have a 30 degree phase difference relative to the signal of the CLRH line with the empty reservoir. In that manner, we will set up the phase detector to operate in the linear regime and to measure a maximum of 120 degrees of the phase difference for different fluids in the channel (i.e., from air to water for permittivity in the range between 1 and 80.1). The permittivity of different fluids used in simulations corresponds to the real permittivity at the operating frequencies of different fluids such as oil, isopropanol, methanol, ethanol, and so on (and is used later for validation).

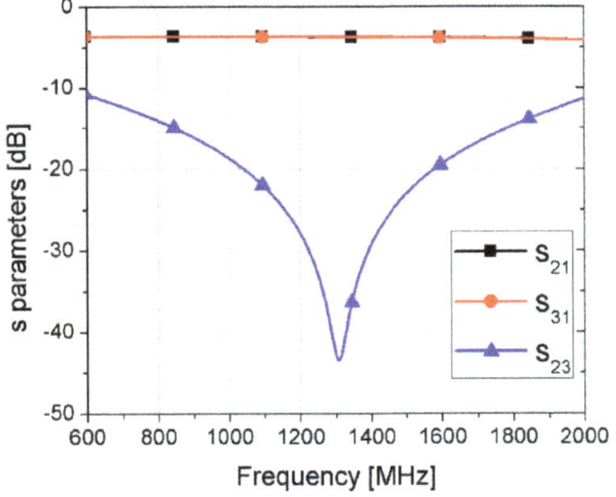

Figure 5. Simulated response of the Wilkinson power divider.

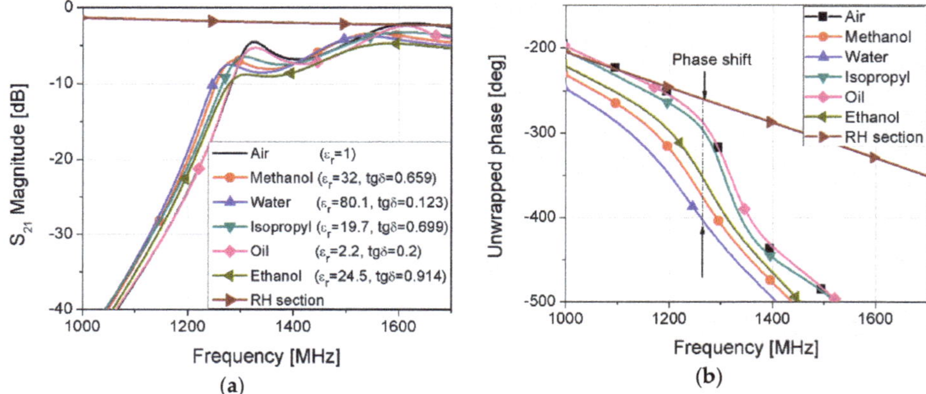

Figure 6. Simulated transmission responses of the RH TL section and CLRH sensing section for different fluids in the microfluidic reservoir: (**a**) magnitude of the transmission characteristics; (**b**) normalized unwrapped phase. The values of the dielectric permittivity (ε_r) and dielectric loss tangent ($tg\delta$) used in simulations are shown in brackets.

The designed divider and TLs are fabricated together with a microfluidic reservoir onto the combined PVC and PMMA substrate, as was described in Section 2. The fabricated prototype of the TL sections with the integrated Wilkinson divider is shown in Figure 7.

Figure 7. Prototype of the proposed sensor with the integrated Wilkinson power divider.

The phase detector module used for measuring the phase shift between two signals propagating thought RH and CLRH TLs is made using integrated circuit AD8302 Analogue Devices (Figure 8a). The phase detector circuit developed to operate in the linear regime is set to the phase difference measurement mode according to the manufacturer's recommendation. Integrated circuit AD8302 on its output provides a voltage signal proportional to the phase difference of the signals on its inputs. A vector network analyzer (VNA) was used for experimental verification and accuracy testing of the designed phase-shifter device. The fabricated circuit of the phase detector manufactured using standard PCB technology is shown in Figure 8b.

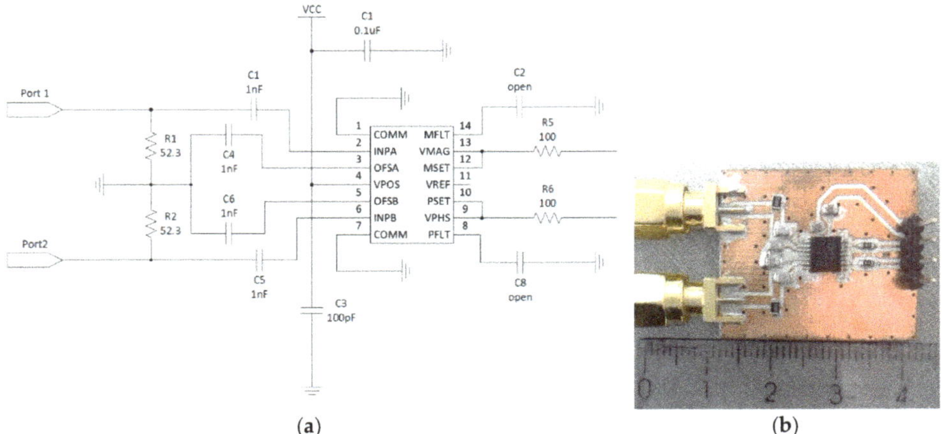

(a)   (b)

Figure 8. Phase detector: (**a**) detailed electronic circuit and (**b**) fabricated circuit.

4. Results

The measurement setup is shown in Figure 9. It consists of a VNA (which is used for characterization of the TLs and generation of the input signal), proposed sensors with the power divider, the phase detector, and the MSP430FR6989 LaunchPad Development Kit [28] for measuring the output voltage and the syringe pump.

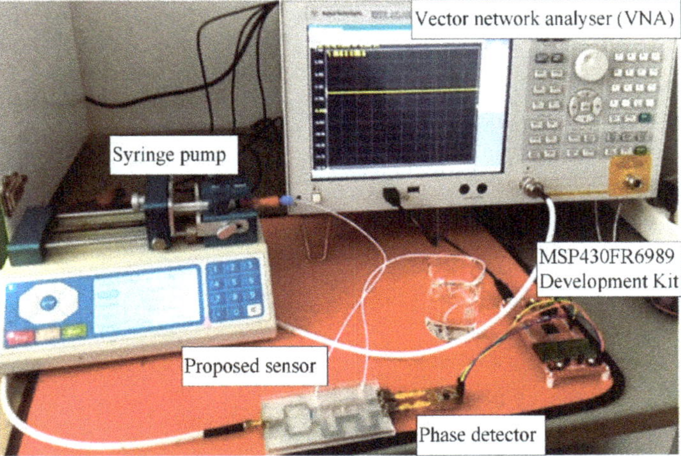

Figure 9. Photograph of the measurement setup.

Firstly, the designed sensor was characterized using a three-port measurement of the transmission signal. Figure 10a shows the transmission characteristics of the RH and LH sections in the frequency range of interest for air and water placed in the microfluidic reservoir and a comparison with the results of the simulation. Good agreement was obtained between the measured and simulated results, except for the small shift in the frequency. It should be noted that the measured amplitude of the signals was several dB larger than in the simulations. However, the maximum difference between the signal through the CLRH and RH sections was still within the 10 dB range. Increased losses arose as a result of imperfections in the fabrication process and were also related to the low conductivity of the aluminum tape used for the conductive layers. Due to the small frequency shift, the phase difference was reduced for 9.5% and the maximal detected phase shift was 112.3 degrees.

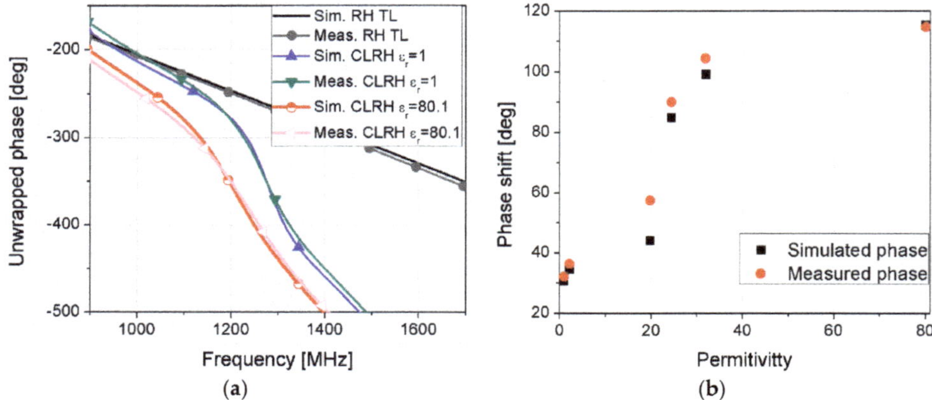

Figure 10. Comparison between simulated and measured phase responses: (**a**) phase responses for air ($\varepsilon_r = 1$) and water ($\varepsilon_r = 80.1$) inside the microfluidic reservoir; (**b**) phase shift versus dielectric constant in the microfluidic reservoir for different fluids.

In the next step, the phase detector was used at the output to measure the phase difference for different fluids inside the microfluidic reservoir. Figure 11 shows the variation of the output voltage versus the dielectric constant of the fluid in the reservoir. The output voltage almost linearly depended on the dielectric constant of the fluid in the reservoir with a regression factor of $R^2 = 0.94$. The limit of detection of the output voltage of the proposed device was determined by the accuracy of the AD8302 integrated circuit and was equal to 10 mV. It can be mentioned that the imaginary part of the complex permittivity does not affect the phase shift, but can influence the amplitude of the signal and insertion losses [1]. The imaginary part of the complex permittivity cannot be directly calculated from the phase shift and therefore the measurement of the amplitudes of two signals should be taken into account. Since the used phase comparator allows a measurement of the amplitude difference between two signals, it can be used for estimating the imaginary part of the complex permittivity, or for mitigation of the comparator measurement error of the phase shift.

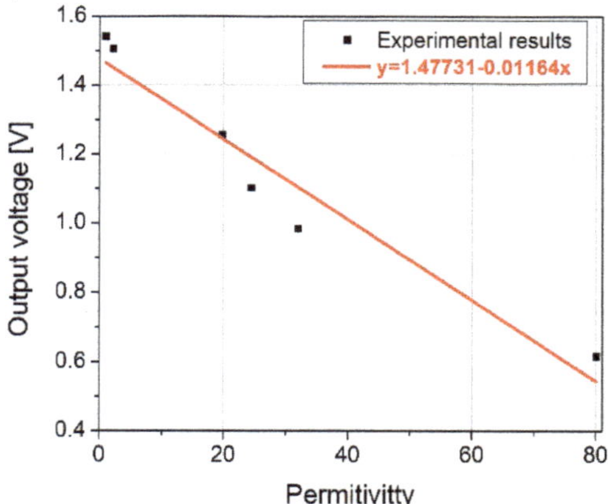

Figure 11. Output voltage as a function of the dielectric constant in the reservoir.

The proposed sensor was used to measure the cell concentration (i.e., biomass) inside the microfluidic bioreactor. MRC-5 human fibroblasts were grown in DMEM with 4.5% glucose supplemented with 10% fetal calf serum and antibiotic/antimycotic solution. The cell density (number of cells per unit volume) and the percentage of viable cells were determined before the measurement using the proposed sensor. The measurement by the sensor was performed immediately after seeding the cells into the reservoir while still free-floating in suspension. This is done for proof-of-concept purposes. Figure 12 shows the output voltage as a function of the number of cells in 1 mL of medium solution. The number of cells influences the effective permittivity of the fluid in the reservoir. Therefore, the effective permittivity of the substrate under the CLRH TL changes and consequently the output voltage changes. Although the change in output voltage is relatively small due to the high permittivity of the medium solution, the measurement response possesses better linearity in terms of the variation of the number of cells with a regression factor of $R^2 = 0.98$.

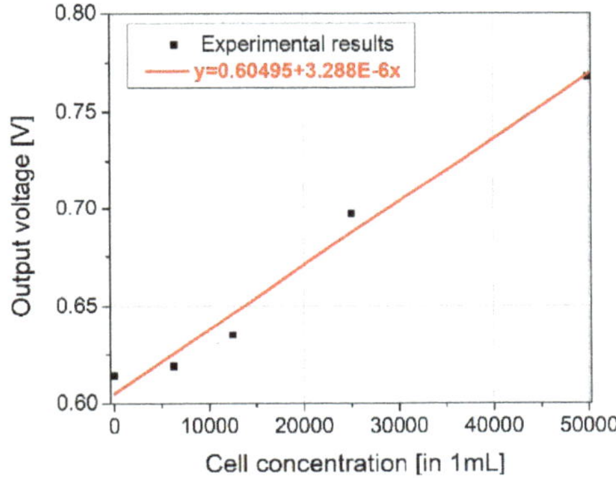

Figure 12. Output voltage as a function of the cell concentration in the microfluidic bioreactor.

5. Discussion

The metamaterial CLRH transmission line approach was used to make a novel microfluidic sensor for the characterization of fluid flowing in the microfluidic reservoir. The proposed sensor comprises a power divider, microstrip lines, and a phase detector designed as a low-cost microfluidic platform using hybrid fabrication technology combining laser micromachining process, xurography, and lamination techniques. The CLRH TL fabricated above the channel is used to improve the sensitivity of the conventional TL in the phase comparator. The TL method was used, since it represents a fast and simple method for determination of the dielectric properties of the material, as well as allowing characterization at a single frequency with a high degree of integration with the sensor elements. The complete read-out detection circuit for determination of permittivity based on the phase shift is suitable for in-field measurement, and has been designed to operate at a frequency of 1.275 GHz.

The measurement results of the fabricated sensor confirm that a change in the permittivity of fluids in the micro reservoir from 1 to 80.1 (from air to water) results in a phase shift of almost 115 degrees. Using the phase advance phenomena of the CLRH line, the sensitivity of the sensor can be improved more than 10-fold compared with the conventional RH line with the same length. The phase difference for the same sensor realized with the RH TL instead of the CLRH one is only 12 degrees. In addition, the designed read-out detection circuit is relatively simple and allows measurement at a single frequency. The output voltage is approaching the linear function of the dielectric constant in the microfluidic reservoir with a regression factor of $R^2 = 0.94$.

Proof of concept for the potential application was demonstrated by implementing the proposed sensor for measurement of the cell concentration in a suspension-like cell culture in a microfluidic bioreactor. The experimental test confirmed that phase difference linearly changes with cell concentration. The signal was relatively low due to the very high dielectric constant of the cell suspension (i.e., the cells floating in the medium). However, the sensor was able to detect even such a low signal intensity.

The proposed device based on a metamaterials-based CLRH sensor presents a low-cost detection solution characterized by relatively high sensitivity and linearity, and therefore it can be used for monitoring small concentrations of specific fluids in different mixtures. On the other hand, the proposed sensor is suitable for a number of biomedical applications utilizing suspension cell cultures, or for fluid characterization.

Author Contributions: V.R. proposed the idea, carried out the experiments and data analysis, and prepared the original drafts. G.K. contributed ideas, methodology for characterization and testing, and developed the phase detector. S.B. and I.P. performed simulations and manufactured the sensor, performed validation, as well as contributed to writing. M.D. and I.G.S. performed experimental verification with cell culture and microfluidic bioreactor and contributed to the manuscript review and editing.

Funding: This work was funded in the framework of project III66004, Development of new information and communication technologies, based on advanced mathematical methods, with applications in medicine, telecommunications, power systems, protection of national heritage and education, Ministry of Education, Science and Technological Development (R Serbia); and REALSENSE1: Monitoring of cell culture parameters using sensors for biomass and nutrients/metabolites in media: Lab-on-a-Chip (LOC) approach, Good Food Institute 2018 Competitive Grant Program.

Conflicts of Interest: The authors declare no conflict of interest.

References

1. Pendry, J.B.; Holden, A.J.; Stewart, W.J.; Youngs, I. Extremely Low Frequency Plasmons in Metallic Mesostructures. *Phys. Rev. Lett.* **1996**, *76*, 4773–4776. [CrossRef] [PubMed]
2. Pendry, J.B.; Holden, A.J.; Robbins, D.J.; Stewart, W.J. Magnetism from conductors and enhanced nonlinear phenomena. *IEEE Trans. Microw. Theory Tech.* **1999**, *47*, 2075–2084. [CrossRef]
3. Anselmi, N.; Gottardi, G. Recent Advances and Current Trends in Metamaterial-by-Design. *J. Phys. Conf. Ser.* **2018**, *963*, 012011. [CrossRef]
4. Laudato, M.; Barchiesi, E. *Wave Dynamics, Mechanics and Physics of Microstructured Metamaterials*, 1st ed.; Springer: Bolingbrook, IL, USA, 2018. [CrossRef]
5. Tong, X.C. *Functional Metamaterials and Metadevices*; Springer: Bolingbrook, IL, USA, 2018. [CrossRef]
6. Available online: https://www.marketwatch.com/press-release/upcoming-trends-of-metamaterials-market-2019-covers-industry-share-size-gross-margin-future-trends-demand-business-insight-by-leading-key-players-forecast-till-2024-2019-06-14 (accessed on 31 October 2019).
7. Song, Q.; Zhang, W.; Wu, P.C.; Zhu, W.; Shen, Z.X.; Chong, P.H.J.; Liang, Q.X.; Yang, Z.C.; Hao, Y.L.; Cai, H.; et al. Water-Resonator-Based Metasurface: An Ultrabroadband and Near-Unity Absorption. *Adv. Opt. Mater.* **2017**, *5*, 1–8. [CrossRef]
8. Song, Q.H.; Zhu, W.M.; Wu, P.C.; Zhang, W.; Wu, Q.Y.S.; Teng, J.H.; Shen, Z.X.; Chong, P.H.J.; Liang, Q.X.; Yang, Z.C.; et al. Liquid-metal-based metasurface for terahertz absorption material: Frequency-agile and wide-angle. *APL Mater.* **2017**, *5*, 066103. [CrossRef]
9. Vivek, A.; Shambavi, K.; Alex, Z.C. A review: Metamaterial sensors for material characterization. *Sens. Rev.* **2019**, *39*, 417–432. [CrossRef]
10. Zhou, H.; Hu, D.; Yang, C.; Chen, C.; Ji, J.; Chen, M.; Chen, Y.; Yang, Y.; Mu, Z. Multiband sensing for dielectric property of chemical using metamaterial integrated microfluidic sensor. *Sci. Rep.* **2018**, *8*, 14801. [CrossRef] [PubMed]
11. Lai, A.; Caloz, C.; Itoh, T. Composite right/left-handed transmission line metamaterials. *IEEE Microw. Mag.* **2004**, *5*, 34–50. [CrossRef]
12. Baker-Jarvis, J.; Janezic, M.D.; Riddle, B.F.; Johnk, R.T.; Kabos, P.; Holloway, C.L.; Geyer, R.G.; Grosvenor, C.A. *Measuring the Permittivity and Permeability of Lossy Materials: Solids, Liquids, Metals, Building Materials, and Negative-Index Materials*; U.S. Department of Commerce, Technology Administration, National Institute of Standards and Technology: Gaithersburg, MD, USA, 2005.

13. Krraoui, H.; Mejri, T.; Aguili, T. Dielectric constant measurement of materials by a microwave technique: Application to the characterization of vegetation leaves. *J. Electromagn. Waves Appl.* **2016**, *30*, 1643–1660. [CrossRef]

14. Brodie, G.; Jacob, M.; Farrell, P. *Microwave and Radio-Frequency Technologies in Agriculture: An Introduction for Agriculturalists and Engineers*, 1st ed.; Sciendo: Warsaw, Poland, 2015.

15. Nozaki, R.; Bose, T.K. Broadband Complex Permittivity Measurements by Time-Domain Spectroscopy. *IEEE Trans. Instrum. Meas.* **1990**, *39*, 945–951. [CrossRef]

16. Cataldo, A.; Tarricone, L.; Attivissimo, F.; Trotta, A. A TDR method for real-time monitoring of liquids. *IEEE Trans. Instrum. Meas.* **2007**, *56*, 1616–1625. [CrossRef]

17. Basics of Measuring the Dielectric Properties of Materials. Available online: https://www.cmc.ca/wp-content/uploads/2019/08/Basics_Of_MeasuringDielectrics_5989-2589EN.pdf (accessed on 31 October 2109).

18. Stuchly, M.A.; Stuchly, S.S. Coaxial Line Reflection Methods for Measuring Dielectric Properties of Biological Substances at Radio and Microwave Frequencies—A Review. *IEEE Trans. Instrum. Meas.* **1980**, *29*, 176–183. [CrossRef]

19. Chen, L.F.; Ong, C.K.; Neo, C.P.; Varadan, V. *Microwave Electronics: Measurement and Materials Characterization*; John Wiley & Sons, Ltd.: Hoboken, NJ, USA, 2004. [CrossRef]

20. Yeow You, K. Effects of Sample Thickness for Dielectric Measurements Using Transmission Phase-Shift Method. *Int. J. Adv. Microw. Technol.* **2016**, *1*, 64–67.

21. Kitic, G. *Microwave Soil Moisture Sensor Based on Phase Shift Method Independent of Electrical Conductivity of the Soil*; The Intellectual Property Office Republic of Serbia: Beograd, Serbia, 2018.

22. Radonić, V.; Birgermajer, S.; Kitić, G. Microfluidic EBG sensor based on phase-shift method realized using 3D printing technology. *Sensors* **2017**, *17*, 892. [CrossRef] [PubMed]

23. Kojic, S.P.; Stojanovic, G.M.; Radonic, V. Novel cost-effective microfluidic chip based on hybrid fabrication and its comprehensive characterization. *Sensors* **2019**, *19*, 1719. [CrossRef] [PubMed]

24. Analog Devices Data Sheet. Available online: http://www.analog.com/media/en/technical-documentation/data-sheets/ad8302.pdf (accessed on 31 October 2019).

25. Sonnet Software. Available online: http://www.sonnetsoftware.com/ (accessed on 31 October 2109).

26. CST Studio Suite, Electromagnetic Filed Simulation Software. Available online: https://www.3ds.com/products-services/simulia/products/cst-studio-suite/ (accessed on 31 October 2109).

27. Radonic, V.; Cselyuszka, N.; Crnojevic-Benging, V.; Kitic, G. Phase-Shift Transmission Line Method for Permittivity Measurement and Its Potential in Sensor Applications. In *Electromagnetic Materials and Devices*; Intech Open: Rijeka, Croatia, 2018. [CrossRef]

28. MSP430FR6989 LaunchPad™ Development Kit (MSP-EXP430FR6989). User's Guide; Texas Instruments. 2015. Available online: http://www.ti.com/lit/ug/slau627a/slau627a.pdf (accessed on 31 October 2109).

 *electronics*

Article

Identifying Near-Perfect Tunneling in Discrete Metamaterial Loaded Waveguides

Kimberley W. Eccleston * and Ian G. Platt

Lincoln Agritech Ltd., Canterbury 7640, New Zealand; ian.platt@lincolnagritech.co.nz
* Correspondence: kim.eccleston@lincolnagritech.co.nz; Tel.: +64-3-325-3741

Received: 25 October 2018; Accepted: 3 January 2019; Published: 11 January 2019

Abstract: Mu-negative and epsilon-negative loaded waveguides taken on their own are nominally cut-off. In ideal circumstances, and when paired in the correct proportions, tunneling will occur. However, due to losses and constraints imposed by finite-sized constituent elements, the ability to experimentally demonstrate tunneling may be hindered. A tunnel identification method has been developed and demonstrated to reveal tunneling behavior that is otherwise obscured. Using ABCD (voltage-current transmission) matrix formulation, the S-parameters of the mu-negative/epsilon-negative loaded waveguide junction is combined with S-parameters of an epsilon-negative loaded waveguide. The method yields symmetric scattering matrices, which allows the effect of losses to be removed to provide yet clearer identification of tunneling.

Keywords: evanescent field tunneling; metamaterials; mu-negative material; epsilon-negative material; split-ring-resonators; waveguides

---

1. Introduction

The purpose of this work is to identify tunneling across an experimental microwave metamaterial heterojunction that may be obscured due to losses and constraints imposed by finite-sized constituent elements in real microwave metamaterials.

Microwave metamaterials exhibit unusual electromagnetic properties (such as negative permittivity and negative permeability) which yield a range of unusual phenomena. One such phenomenon is the tunneling of electromagnetic waves through epsilon-near-zero (ENZ) metamaterial [1–5], epsilon-negative (ENG) metamaterial (or empty waveguides operated at cut-off) paired with mu-negative (MNG) metamaterial [6–15], and cut-off waveguides either filled [16,17], or lined [18,19], with metamaterials. In this work, we are specifically interested in tunneling in a microwave waveguide filled by a ENG/MNG metamaterial heterojunction where the waveguide is cut-off when loaded by either the ENG or MNG metamaterial on its own [7,14].

ENG behavior can be obtained at microwave frequencies using an empty cut-off waveguide [7,14] or by loading the waveguide with an array of thin wires [20,21]. MNG behavior can be obtained at microwave frequencies by loading the waveguide with an array of split-ring resonators (SRRs) [7,14,20,22]. In both cases, lumped elements can also be used [8–10,13]. Practical microwave metamaterials are constructed from a discrete number of finite-sized elements. Therefore, in a practical experiment, it may not be possible to satisfy the length requirement (or attenuation condition) for perfect tunneling [6,7]. Hence, it is the purpose of this work to develop a method to identify tunneling behavior in discrete microwave ENG/MNG metamaterial junctions.

Whether using loaded waveguides or transmission lines, resistive losses impact on the ability to demonstrate tunneling [6,15]. Therefore, an additional purpose of this work is to isolate losses from experimental results to further reveal tunneling.

2. Theoretical Background

In this work we confine ourselves to time-harmonic electromagnetic waves. At a given angular frequency ω, each field component of an electromagnetic wave travelling in the positive z direction will contain the factor $\exp(j\omega t - \gamma z)$ where γ is the propagation constant and $j = \sqrt{-1}$. In general γ is complex valued and related to real valued attenuation and phase constants, α and β respectively, taking the form $\gamma = \alpha + j\beta$. For passive media, the amplitude of the field components will either decay or remain constant with z and this means α is always greater than or equal zero. For the rest of this work we will drop explicit reference to the factor $\exp(j\omega t - \gamma z)$ and describe such quantities using phasor notation. Within the waveguide, we confine ourselves to transverse-electric (TE) wave propagation and consider the lowest order mode being the TE_{10} mode [23] since this simplifies the procedure to demonstrate tunneling.

Figure 1 shows a diagram of an idealized waveguide tunneling configuration which comprises four sections: Sections 1 and 4 air-filled, and Sections 2 and 3 being filled by μ-negative (MNG) and ε-negative (ENG) metamaterials respectively. For the moment, it will be assumed that the MNG and ENG metamaterials are continuous, lossless and isotropic. The waveguide cross-section is identical along the structure so the only discontinuity is a change of material along the longitudinal axis (z).

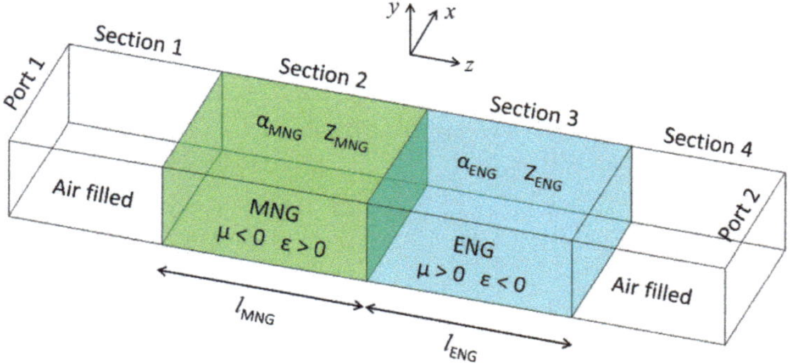

Figure 1. Schematic of a cascade of MNG and ENG metamaterial filled waveguide where tunneling (transmission from port 1 to port 2 or vice versa) is possible under certain conditions. (In color.).

At a frequency above the TE_{10} mode cut-off frequency of Sections 1 and 4, Sections 1 and 4 will support propagation. On the other hand, Sections 2 and 3 on their own, because their permittivity and permeability are opposite signed, will be cut-off (see Appendix A). It can be shown (see Appendix B) perfect transmission from port 1 to port 2 (zero insertion loss and zero reflection coefficient) occurs when [6,7]:

$$\alpha_{ENG} l_{ENG} = \alpha_{MNG} l_{MNG} \tag{1}$$

and:

$$Z_{ENG} = -Z_{MNG} \tag{2}$$

where l_{MNG} and l_{ENG} are the lengths of the MNG and ENG loaded waveguides respectively, and Z_{MNG} and Z_{ENG} are the TE_{10} mode wave-impedances of the MNG and ENG metamaterials respectively. We denote (1) and (2) the attenuation and impedance tunneling conditions respectively.

As only evanescent modes can be excited in waveguides filled with lossless isotropic MNG and ENG metamaterials, Z_{MNG} and Z_{ENG} will be purely imaginary with $\mathrm{Im}(Z_{MNG}) < 0$ and $\mathrm{Im}(Z_{ENG}) > 0$ (see Appendix A). Provided Z_{MNG} and Z_{ENG} are of similar magnitude, the impedance tunneling condition can be satisfied at some frequency. On the other hand, the attenuation constant is always positive valued and this requires a certain length ratio to satisfy the attenuation tunneling condition.

Similar conclusions are obtained for waveguides filled with certain continuous lossless anisotropic MNG and ENG metamaterials (see Appendix C) [7].

The single ENG/MNG heterojunction contained within the waveguide of Figure 1 represents the minimum configuration for tunneling. A five section waveguide tunneling structure may be realized as shown in Figure 2 where the length of Section 3 (ENG) is $2l_{ENG}$. The tunneling conditions of Figure 2 are the same as Figure 1, and hence equations (1) and (2) apply to Figure 2. It is the structure of Figure 2 which forms the theoretical basis for the experimental tunneling identification. This structure is symmetrical which is useful when dealing with losses.

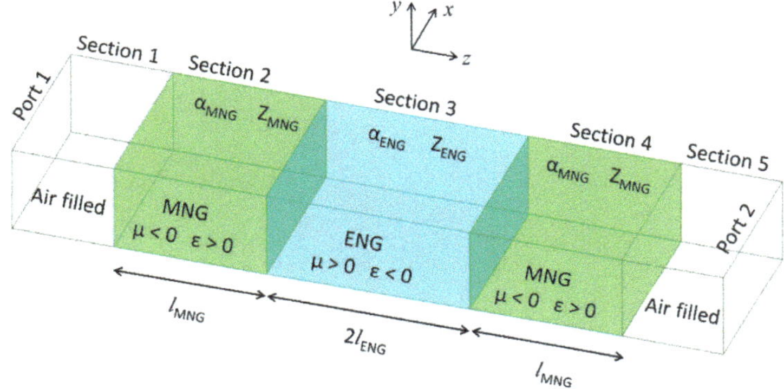

Figure 2. Schematic of a cascade of MNG, ENG and MNG metamaterial filled waveguide where tunneling (transmission from port 1 to port 2 or vice versa) is possible under certain conditions. (In color.).

3. Tunneling in Practical Microwave Metamaterials

Practical microwave metamaterials employing split-ring resonators (SRRs) and thin wires to obtain MNG and ENG behavior respectively, are crystalline (periodic) structures with lattice constants of the order of millimeter [7,14,16,17,20,24,25]. So whilst significantly smaller than the guide wavelength, thereby approximating the continuous medium, the dimensions of practical metamaterial elements are never-the-less finite and means that material lengths take discrete values. It is unlikely that the attenuation tunneling condition is satisfied at the same frequency as the impedance tunneling condition when the ENG and MNG lengths are restricted to discrete values.

For example, for ENG and MNG metamaterials based upon the negative-refractive-index (NRI) metamaterials described by Eccleston & Platt [24], the length of the MNG metamaterial would be integer multiples of 5 mm and the ENG metamaterial takes lengths in integer increments of 20 mm. The significant difference in lattice constants arises from the need to achieve useful permittivity values within the low gigahertz range using an entirely printed structure [24]. As will be demonstrated in Section 5, perfect tunneling in such materials is obtained at 2.73 GHz when l_{MNG} = 25 mm and l_{ENG} = 8.8 mm, which is incompatible with these lattice constraints of the materials. At higher frequencies, the lattice constant for ENG could be comparable to that of the MNG [25].

On the other hand, this problem does not arise when one of the metamaterials is continuous. For example, Baena et al. [7] use a SRR waveguide and an empty cut-off waveguide for MNG and ENG behavior respectively; l_{MNG} and l_{ENG} are discrete and continuous respectively.

4. Tunnel Identification Principle

The procedure that will be described is aimed at identifying tunneling in a MNG/ENG junction constructed entirely from discrete metamaterials. Two sets of microwave scattering parameters [23] is required: (i) ENG/MNG filled waveguide junction, and (ii) ENG filled waveguide of known length.

The former structure will be denoted the device-under-test (DUT) and contains the MNG/ENG heterojunction required for tunneling.

It will be assumed that MNG behavior is obtained over a narrow bandwidth using an array of split-ring resonators (SRRs), and ENG behavior is obtained over a significantly wider bandwidth using an array of strips. The term strip is used rather than wire to acknowledge implementation in printed-circuit-board (PCB) or similar technology [24,25].

The block diagram describing tunneling identification is shown in Figure 3a. The DUT and reversed DUT blocks represent the direct measurements of the DUT, whilst the middle section, of length l_{Strip2}, is described by a transmission line model. The propagation constant and wave-impedance of the transmission line model are extracted from measurements of the strip loaded (ENG) waveguide. The length l_{Strip2} of the middle section of Figure 3a is in general different to the length of the measured strip loaded waveguide.

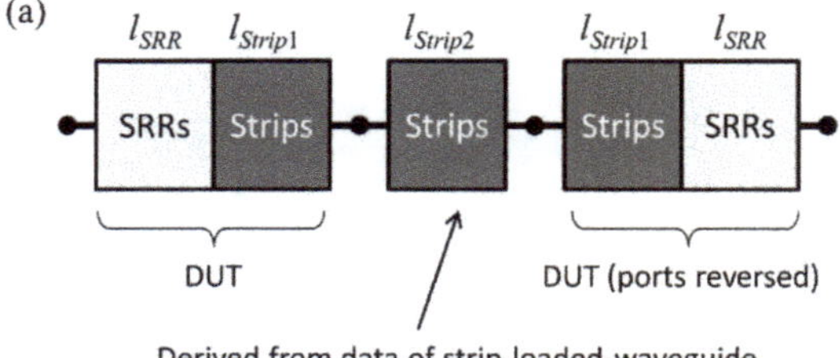

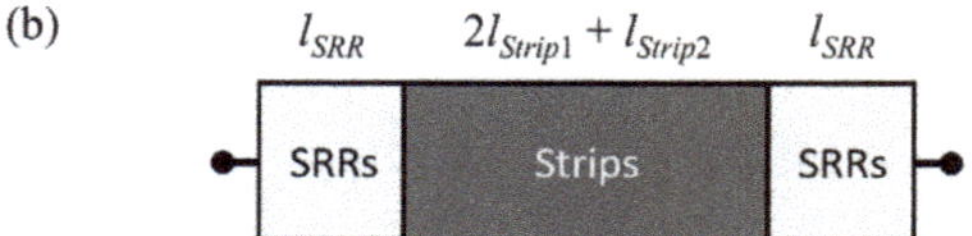

Figure 3. Block diagram of the tunneling identification method: (**a**) constituent elements, and (**b**) equivalent structure.

The cascade of Figure 3a is equivalent to the block diagram shown in Figure 3b. Comparing Figure 3b with Figure 2, then applying (1) and (2) to Figure 3b, it can be shown for lossless media, perfect tunneling occurs when:

$$2\alpha_{SRR}l_{SRR} = \alpha_{Strip}\left(2l_{Strip1} + l_{Strip2}\right) \tag{3}$$

and

$$Z_{SRR} = -Z_{Strip} \tag{4}$$

The ability to apply Figure 3a and its equivalent Figure 3b assumes only the dominant mode exists at the ports of each waveguide section depicted in Figure 3a, and that the strip loaded waveguide behaves as a homogenously filled waveguide. Provided that the waveguide transverse cross section is the same for all sections, the method does not introduce new discontinuities; the only discontinuity being the essential MNG/ENG (SRR/Strip) interface inherently contained in the DUT.

The block diagram of Figure 3a can be analyzed using ABCD (voltage-current transmission) parameters which are related to scattering parameters [26]. The ABCD matrix, $\mathbf{A}_a$, (containing the ABCD parameters) of the circuit of Figure 3a is given by:

$$\mathbf{A}_a = \mathbf{A}_{DUT}\mathbf{A}_{Strip}(l_{Strip2})\bar{\mathbf{A}}_{DUT} \tag{5}$$

where $\mathbf{A}_{DUT}$ and $\mathbf{A}_{Strip}(l_{Strip2})$ are the ABCD matrices of the DUT and strip loaded waveguide of length l_{Strip2} respectively, and $\overline{\mathbf{A}}_{DUT}$ is the ABCD matrix of the DUT in the reverse direction and is related to the elements of $\mathbf{A}_{DUT}$ by [27]:

$$\overline{\mathbf{A}}_{DUT} = \begin{bmatrix} D_{DUT} & B_{DUT} \\ C_{DUT} & A_{DUT} \end{bmatrix} \tag{6}$$

where

$$\mathbf{A}_{DUT} = \begin{bmatrix} A_{DUT} & B_{DUT} \\ C_{DUT} & D_{DUT} \end{bmatrix} \tag{7}$$

Applying the continuous material approximation for the strip loaded waveguide, a transmission line model for the strip loaded waveguide of length l can be used [23]:

$$\mathbf{A}_{Strip}(l) = \begin{bmatrix} \cosh(\gamma_{Strip}l) & Z_{Strip}\sinh(\gamma_{Strip}l) \\ \dfrac{\sinh(\gamma_{Strip}l)}{Z_{Strip}} & \cosh(\gamma_{Strip}l) \end{bmatrix} \tag{8}$$

where γ_{Strip} is its complex propagation constant and Z_{Strip} is its wave-impedance. Both γ_{Strip} and Z_{Strip} are obtained from S-parameter measurements of a known length of strip loaded waveguide. This is achieved by relating (8) to the ABCD matrix of the measured strip loaded waveguide of known length l.

Conversion between S-parameters and ABCD parameters requires knowledge of the port reference impedance [26]. By using the same waveguide transverse cross section for both the DUT and strip loaded waveguide, the ambiguity associated with waveguide characteristic impedance [28], and port reference impedance is avoided. Hence, the port reference impedance used here is taken as directly proportional to the TE_{10} mode wave-impedance of an empty waveguide.

The ABCD matrix $\mathbf{A}_a$ is calculated using (5) over a range of frequencies and l_{Strip2}. The overall scattering matrix of Figure 3b can be obtained from $\mathbf{A}_a$. Therefore, $|S_{21}|$ can be plotted as a function of frequency and l_{Strip2} from which a suitable value of l_{Strip2}, as well as the tunneling frequency can be identified. Negative values of l_{Strip2} are permissible, but only if the total length of the ENG sections, $(2l_{Strip1} + l_{Strip2})$, is positive and non-zero. Losses will limit the maximum transmission that can be achieved.

The complimentary method, which uses data from a SRR loaded waveguide can be used to identify tunneling in the DUT. Similarly, a method that uses data from both a SRR loaded waveguide and strip loaded waveguide can also be developed. Neither of these methods will be considered here, as losses as well as the highly resonant behavior of SRRs renders these approaches unfeasible.

5. Description and Theoretical Analysis of Experimental Metamaterials

The purpose of this section is to show that tunneling is theoretically possible in accordance to the mechanism described in Section 2, for strip loaded and SRR loaded waveguides described in [24].

5.1. Description of Metamaterials

Figure 4 gives a rendering of the DUT (MNG and ENG loaded waveguide junction) of which one layer of thickness 5 mm (or one period of 5 mm in the y direction) is shown for clarity. To fill a WR284 waveguide used in the experiment, 7 layers (or 7 periods in the y direction) are required. Also shown are 72 mm wide waveguide feeds at either end which correspond to WR284 waveguide feeds used in the experiment. Both the SRR and strip loaded regions have a length of 25 mm and a width of 65 mm.

The MNG region comprises an array of SRRs aligned transverse to the waveguide to achieve negative permeability in the x-direction and will interact with the x component of magnetic field of the TE_{10} mode. The ENG region comprises y-directed strips to achieve negative permittivity in the y-direction thereby interacting with the TE_{10} mode electric field. This arrangement gives anisotropic permeability and permittivity [7,24] and field analysis given in Appendix C shows that tunneling described in Section 2 is indeed possible for such metamaterial junctions. The SRRs and strips are mounted on 13 printed circuit boards (PCBs) mounted parallel to the yz plane and are spaced 5 mm in the x direction to span a width of 65 mm. On the PCBs containing the SRRs, the SRRs have 5 mm periodicity in both directions and this means that within the SRR loaded waveguide, the SRRs have three-dimensional periodicity of 5 mm.

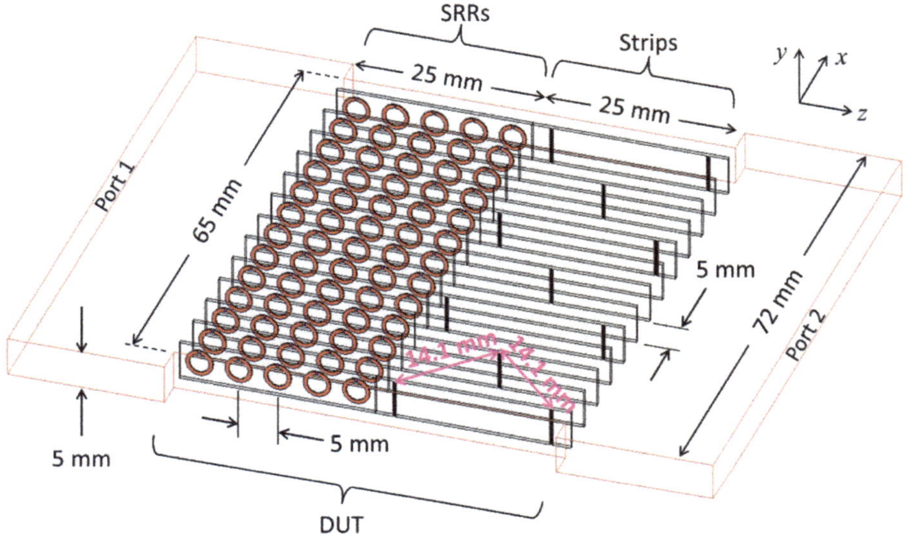

Figure 4. Rendering of one layer of the DUT containing the SRR array (MNG)/strip array (ENG) interface. (In color.).

The design approach and the structural parameters of the SRRs and strips are described in detail elsewhere [24], but the key features are: (i) broadside-coupled SRRs were used as they are compact and minimize bianisotropy [29], (ii) the SRRs were designed to resonate around 3 GHz, (iii) uses a thinned array of strips to achieve suitable values of permittivity [24]. The strip locations have a two-dimensional periodicity of 14.1 mm in the xz plane, and their locations are coincident with the PCBs.

5.2. Theoretical Predictions

Theoretical models for the SRR and strip loaded waveguides are described in earlier work [24] and are used to predict the μ_{XX} and ε_{YY} of the SRR loaded waveguide, and ε_{YY} of the strip loaded waveguide. These models include electric polarization of the SRRs and dielectric loading due to the PCB substrates, and use the continuous medium approximation. The other tensor permeability and permittivity diagonal elements are equal to that of free-space, and the off-diagonal elements are zero. Figure 5 shows the frequency response of μ_{XX} and ε_{YY} of the SRR loaded waveguide, and ε_{YY} of the strip loaded waveguide over the frequency range 2.5 GHz to 3 GHz. It is apparent that SRR loading provides negative permeability over a narrow band (2.663 GHz to 2.752 GHz) and its value is strongly dependent on frequency. On the other hand, strip loading provides negative permittivity over a much wider frequency range and is a relatively weaker function of frequency.

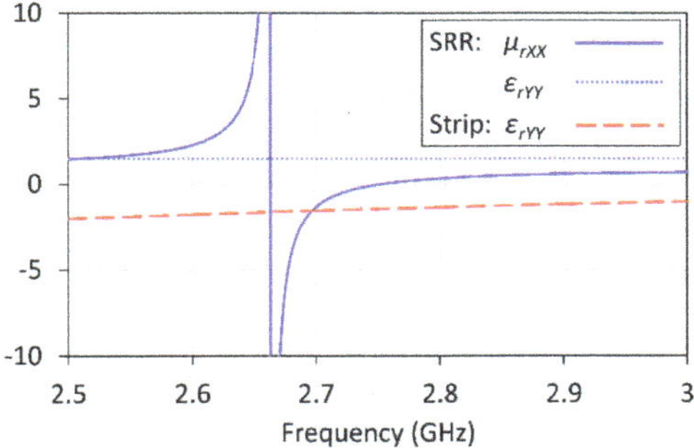

Figure 5. Theoretical predictions of anisotropic permittivity and permeability for the SRR and strip loaded waveguides of the type considered in this work. (In color.).

Using the field analysis of Appendix C, the TE_{10} mode propagation constant and wave-impedance can be calculated from the anisotropic permittivity and permittivity (Figure 5), and are shown in Figure 6. The impedances are normalized to the TE_{10} mode wave-impedance of an empty 65 mm wide waveguide. As the strip loaded waveguide operates in its evanescent mode over the frequency range 2.5 GHz to 3 GHz, its real part of wave-impedance and phase constant are both zero over this range and are not shown. Importantly, the imaginary parts of wave-impedance are opposite signed over the frequency range 2.663 GHz to 2.752 GHz, and hence it is useful to show the sum $\mathrm{Im}\left(Z_{Strip}\right) + \mathrm{Im}\left(Z_{SRR}\right)$ over the range 2.663 GHz to 2.752 GHz in Figure 6a.

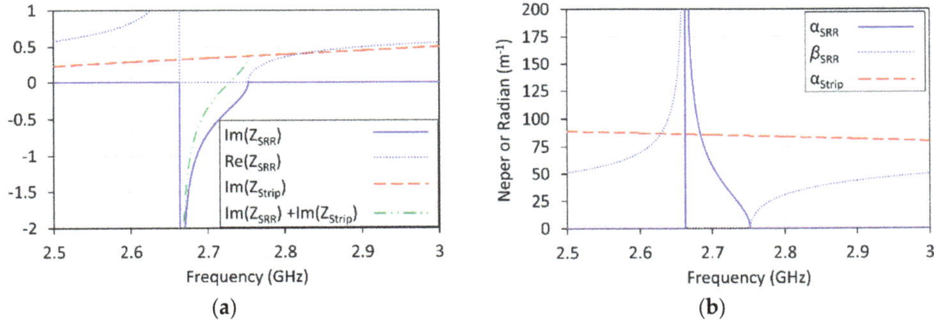

Figure 6. Theoretically calculated TE_{10} mode (**a**) normalized wave-impedance, (**b**) attenuation and phase constants for strip and SRR loaded waveguide. (In color.).

The zero crossing of $\mathrm{Im}\left(Z_{Strip}\right) + \mathrm{Im}\left(Z_{SRR}\right)$ in Figure 6a indicates that the impedance tunneling condition (2) is satisfied at 2.728 GHz. At this frequency, Figure 6b shows that the attenuation constants are 30.1 Np/m and 85.2 Np/m for the SRR and strip loaded waveguide respectively. If the length of the SRR loaded waveguide l_{MNG} is 25 mm, then the strip loaded waveguide length l_{ENG} needs to be 8.9 mm to satisfy the attenuation tunneling condition (1). Figure 7 shows the calculated S-parameter frequency responses for the symmetrical structure of Figure 2 for the two cases (i) l_{MNG} = 25 mm and l_{ENG} = 8.9 mm, and (ii) l_{MNG} = 25 mm and l_{ENG} = 25 mm. It is apparent that for the former case, reflection-less transmission is obtained at 2.728 GHz. On the hand, the latter cases shows low coupling due to the non-satisfaction of the attenuation tunneling condition.

For the case of l_{ENG} equal to 8.9 mm, a very sharp peak in transmission is observed at 2.66 GHz. This frequency falls slightly outside the negative permeability band, and therefore, the SRR loaded waveguides behave as electrically long, high impedance transmission lines (see Figure 6). Due to the high electrical lengths in this vicinity, there will be a frequency (or multiple frequencies) where the length of the SRR loaded waveguides are integer multiples of a half guide-wavelength. In this situation, the SRR loaded waveguides behave as half-wave transformers effectively providing a direct coupling to the strip loaded waveguide to port 1 and 2. Due to the mismatch between the SRR loaded waveguide and empty waveguide wave-impedances, the resulting peak will be narrow. The total ENG length of Figure 2 is 17.8 mm ($2l_{ENG}$) and its attenuation is constant at 2.66 GHz is 86.3 Np/m. Hence, at 2.66 GHz its total attenuation will be 13 dB which is the level of transmission seen in Figure 7. The phenomenon is also denoted a Fabry-Perot resonance [7].

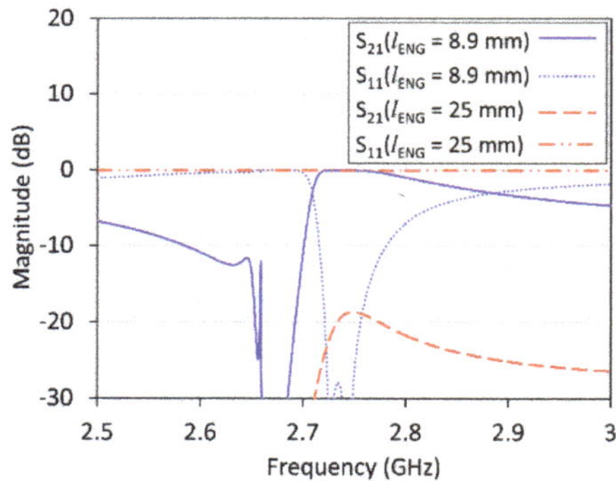

Figure 7. Theoretically predicted TE_{10} mode S-parameters for the structure of Figure 2 using anisotropic permittivity and permeability values of Figure 5 for the two cases of l_{ENG} when l_{MNG} is 25 mm. (In color.).

6. Experiment

6.1. Description of Metamaterials

The metamaterials described in Section 5.1 were fabricated using a multilayer PCB fabrication process on Rogers 4000 series substrates of which individual PCBs were mounted longitudinally in waveguide test fixtures. The PCB thickness was 0.5 mm. The PCBs were 25 mm × 36 mm and each type required to construct the MNG (SRR) and ENG (strip) regions is shown in Figure 8a. The rings of the SRRs use the middle metal layers but are visible through the translucent substrate, whilst the strips are placed on the top metal layer; the bottom metal layer is unutilized. The DUT is 50 mm long ($l_{SRR} = 25$ mm and $l_{Strip1} = 25$ mm). The length of the strip loaded waveguide is 25 mm.

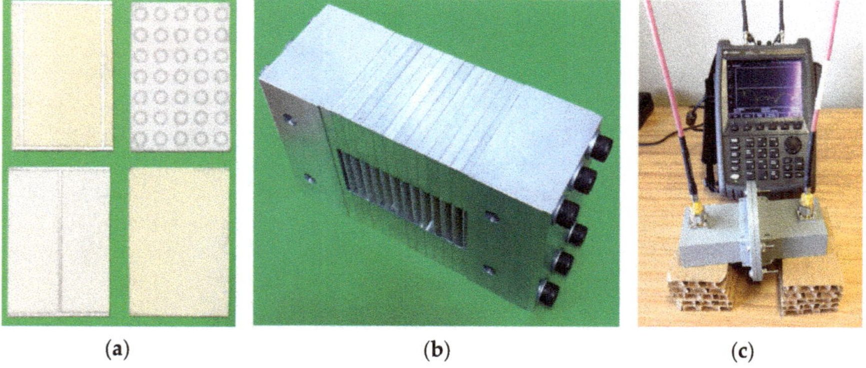

(a)   (b)   (c)

Figure 8. Photo of: (**a**) each PCB type used in the experiment, (**b**) the waveguide PCB mount, and (**c**) measurement equipment shown measuring the *through* configuration for TL deembedding. (In color.).

6.2. Waveguide PCB Mount

Figure 8b shows a photo of the waveguide PCB mount to hold the PCBs associated with the DUT. The one for the strip loaded waveguide is similar but shorter in length. When fully assembled, the waveguide mounts contain 13 longitudinal slots, with width equal to the PCB thickness (0.5 mm), and depth 1 mm, to hold the PCBs with 5 mm spacing across a width of 65 mm. The waveguide PCB mount mates with WR284 (72.14 mm × 34.04 mm) waveguide, and has transverse dimensions of 65 mm × 34 mm.

For manufacturability and to aid inserting PCBs, the waveguide mount is constructed from laminations. Transverse bolts along with the laminated nature of the test fixture, creates a vice to secure the PCBs in the slots. Due to manufacturing tolerances, it is impossible to machine laminations to exactly the same longitudinal length. To ensure continuity of waveguide longitudinal currents (in top and bottom walls) 0.5 mm deep recesses are milled at each end of the mount to create very low impedance RF chokes with the WR284 flanges.

6.3. Measurement Procedure

Measurements were performed over the band 2 GHz to 4 GHz using a Keysight N9918A RF vector network analyzer (Figure 8c) calibrated to SMA coaxial connector reference planes using the short-open-load-thru (SOLT) procedure. To eliminate effects coax-to-waveguide transitions, through-line (TL) deembedding [30] (which is related to TRL deembedding) was conducted offline to obtain the S-parameters for the various loaded waveguide PCB mounts. These measurements are available as Supplementary Materials. The reference planes were then extended by 0.5 mm to remove the effect of the RF chokes at the ends of the waveguide PCB mounts. The S-parameters were finally renormalized from the wave-impedance of an empty 72 mm wide (WR284) waveguide to that of an empty 65 mm wide (waveguide mount) waveguide. As the TE_{10} mode cut-off frequency of the 65 mm wide waveguide is 2.31 GHz, only measurements above this frequency are useful.

6.4. Strip and SRR Loaded Waveguide

Figure 9 shows the attenuation and phase constants, and wave-impedance extracted from the strip loaded waveguide measurements over the range 2.5 GHz to 3.5 GHz. The wave-impedance is normalized to the wave-impedance of an empty 65 mm waveguide. For completeness, the attenuation and phase constants, and wave-impedance for a SRR loaded waveguide is also shown for qualitative purposes. The challenges of retrieval ambiguities [31–35] is not evident for the strip loaded waveguide, whereas it could be a factor for the SRR loaded waveguide at the lower end of the MNG passband.

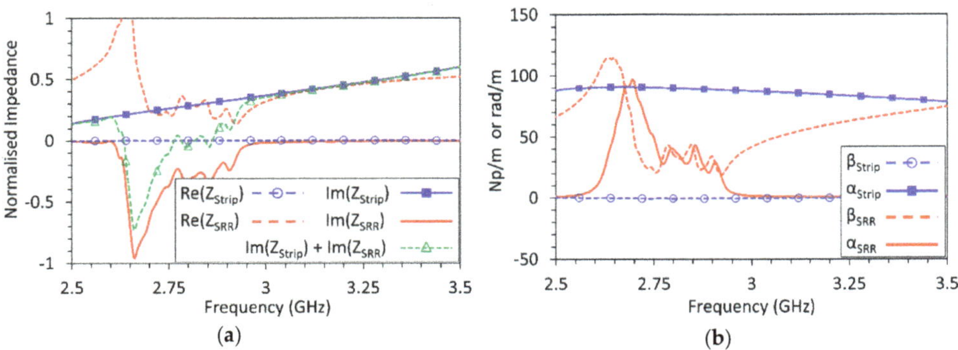

Figure 9. Extracted TE_{10} mode (**a**) normalized wave-impedance, (**b**) attenuation and phase constants for strip and SRR loaded waveguide. (In color.).

Figure 9 shows that the strip loaded waveguide behaves like an ideal evanescent-mode attenuator with an essentially zero phase constant (β) and pure imaginary characteristic impedance over the entire frequency range of interest. The SRR loaded waveguide has almost pure-real wave-impedance and near-zero attenuation constant (α), when operating below about 2.6 GHz and above about 2.95 GHz, and it therefore operates in a propagating mode in these frequency bands. On the other hand, between about 2.65 GHz and 2.95 GHz, it behaves as a lossy evanescent-mode attenuator with complex valued propagation constant and wave-impedance. This lossy behavior is attributed to losses in the SRRs which are operating near resonance [24].

Although it is difficult to make use of the SRR loaded waveguide data directly, it never-the-less can be qualitatively interpreted as evidence of MNG behavior over the range 2.65 GHz to 2.95 GHz. For instance, α_{SRR} and imaginary part of Z_{SRR} are both non-zero, over this range. Figure 9a also shows that the sum of $\mathrm{Im}(Z_{Strip})$ and $\mathrm{Im}(Z_{SRR})$, crosses zero at several frequencies over the range 2.76 GHz to 2.86 GHz; meaning that the impedance tunneling condition (4) will be satisfied in the imaginary part within this range. In an ideal situation, the impedance matching condition (4) is satisfied at one frequency only, so the multiple zeros of $\mathrm{Im}(Z_{Strip}) + \mathrm{Im}(Z_{SRR})$ in Figure 9a is due to SRR resonant frequency variation across the PCB panels [36].

6.5. Tunnel Identification

The tunnel identification method described in Section 4 was applied to the DUT measurements. Figure 10 shows the transmission (S_{21}) as a function of frequency (f) and middle strip loaded waveguide section length (l_{Strip2}). Transmission is maximum at 2.8 GHz, and peaks when l_{Strip2} is -40.8 mm. The l_{Strip2} value of -40.8 mm means that the total length of the equivalent strip loaded section of Figure 3b, $(2l_{Strip1} + l_{Strip2})$, is 9.2 mm. This value is comparable to the theoretical prediction of 17.8 mm given in Section 5.

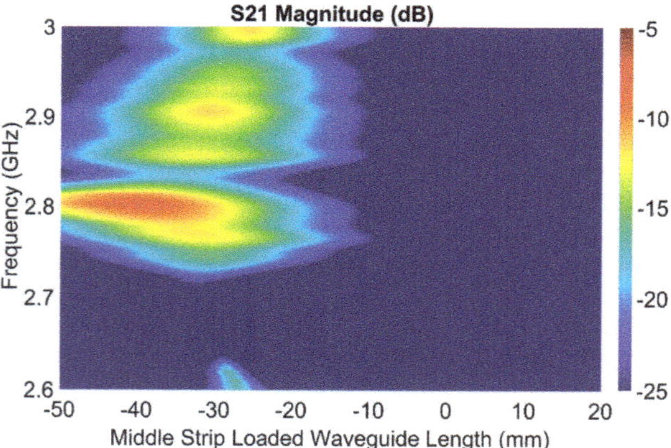

Figure 10. Transmission (S_{21}) magnitude versus frequency and length of the middle strip loaded waveguide (l_{Strip2}). Peak transmission occurs at 2.8 GHz when the length of the middle strip loaded waveguide (l_{Strip2}) is -40.8 mm. (In color.).

Figure 11 shows the S-parameters of the cascaded structure of Figure 3a when l_{Strip2} is equal to -40.8 mm. For reference, the S-parameters of the DUT are shown. At 2.8 GHz, the tunneling effect is clearly revealed with an 8.6 dB boost in transmission (S_{21}) compared to the DUT on its own, as well as in improvement of input and output match. Importantly, the resulting S-parameters are symmetric ($S_{11} = S_{22}$).

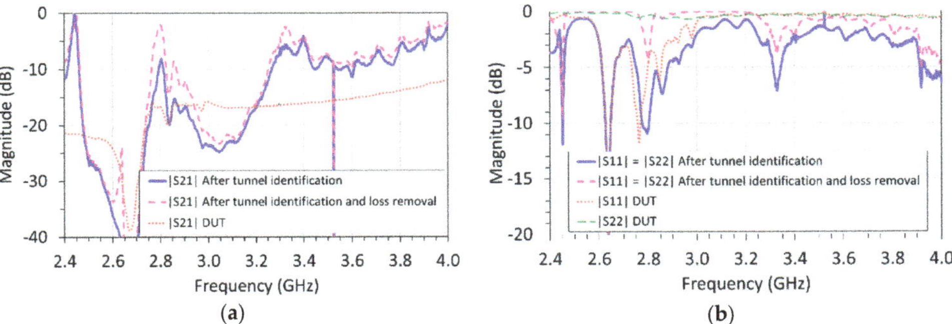

Figure 11. S-parameter magnitude frequency responses after tunnel identification and after loss is removed: (**a**) S_{21}, and (**b**) S_{11} (and S_{22}). The DUT response is included for comparison. (In color.).

For a symmetrical reciprocal two-port, the power loss proportion, L, can be calculated by power conservation:

$$L = 1 - |S_{11}|^2 - |S_{21}|^2 = 1 - |S_{22}|^2 - |S_{12}|^2 \tag{9}$$

where S_{11}, S_{22}, S_{12} and S_{21} are the S-parameters of the two-port. Figure 12 shows the power loss of the DUT after tunnel identification. Significant loss occurs within the MNG band and this explains the insertion loss of 8.1 dB at 2.8 GHz after the tunnel identification process of Section 3. The effects of losses can be removed from the symmetric two-port by scaling the S-parameters by $1/\sqrt{1-L}$. With the effects of losses are removed (pink dashed trace in Figure 11), the resulting tunnel residual insertion loss, due to mismatch, is 2.2 dB.

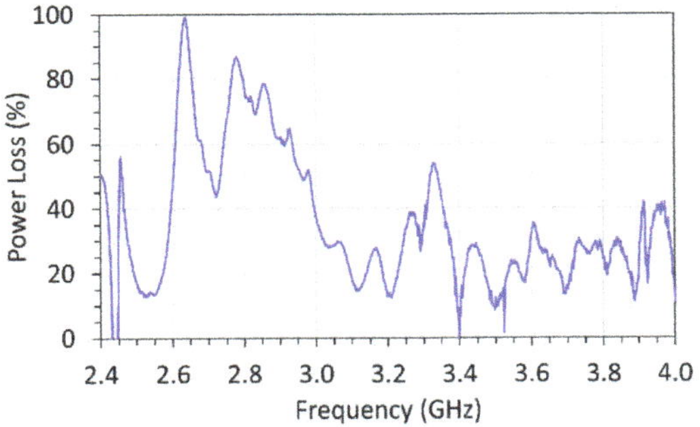

Figure 12. Power loss frequency response for the deembedded DUT.

7. Other Observations

After the tunnel identification method has been applied, a significant transmission boost occurs at 2.45 GHz and 3.3 GHz (blue trace in Figure 11). Both these frequencies fall well outside the MNG regime of the SRR loaded waveguide and therefore cannot be attributed to tunneling. At these frequencies, the SRR loaded waveguide behaves as a transmission line that transform the empty waveguide characteristic impedance to a complex value, and at the same time, the strip loaded waveguide essentially behaves as a lossless evanescent mode attenuator. The strip-loaded waveguide can be treated as a two-port element terminated by complex valued impedances. From microwave amplifier design principles [37], this situation may result in a boost in transmission because of the non-trivial dependence on terminating impedance.

When loss is removed, a sharp peak is also revealed at 2.636 GHz (pink trace in Figure 11a). Figure 9 indicates that this frequency is slightly outside the MNG band of frequencies and therefore, this peak is attributed to Fabry-Perot resonances [7] as discussed in the theoretical investigation in Section 5. This resonance is obscured after tunnel identification (blue trace in Figure 11) as the loss at this frequency (Figure 12), is nearly 100%.

8. Conclusions

Due to the finite size of constituent elements and losses, the ability to experimentally demonstrate tunneling across a cascade of mu-negative and epsilon-negative loaded waveguides may be hindered. A tunnel identification method has been developed and demonstrated to reveal the tunneling behavior that is otherwise obscured. The S-parameters of a mu-negative/epsilon-negative loaded waveguide junction are combined with the S-parameters of an epsilon-negative loaded waveguide. The method yields symmetric S-matrices, which the effect of losses to be removed to provide a yet clearer demonstration of tunneling.

Supplementary Materials: The measured S-parameters, after applying TL deembeding to WR284 waveguide flange reference planes, are available online at http://www.mdpi.com/2079-9292/8/1/84/s1 as .s2p files. In the case of the SRR and Strip loaded waveguide, symmetry is enforced in the deembedding process. In all cases, the reference impedance is the TE_{10} mode wave-impedance of a 72 mm wide waveguide. Therefore, prior to further processing, it is necessary to shift the reference planes 0.5 mm and renormalize the reference impedance of that of a 65 mm waveguide.

Author Contributions: Conceptualization, K.W.E. and I.G.P. Methodology, K.W.E.; Software, K.W.E.; Investigation, K.W.E.; Resources, I.G.P.; Data curation, K.W.E.; Writing—original draft preparation, K.W.E.; Writing—review and editing, K.W.E. and I.G.P.; Project administration, I.G.P.; Funding acquisition, I.G.P.

Funding: This work was funded by the New Zealand Ministry of Business, Innovation and Employment (Research contract no. LVLX1505).

Acknowledgments: The authors would like to thank student intern F. Flamein, of Graduate School of Chemistry, Biology and Physics (ENSCBP): Bordeaux Institute of Technology (Bordeaux INP), for assistance in the conduct of experiments, R Cook of Lincoln Agritech Ltd. for fabricating the waveguide test fixtures, and their colleagues A. Tan and I. M. Woodhead for useful discussions.

Appendix A

In this appendix, the dispersion relations for a lossless isotropic ENG or MNG loaded waveguide are obtained. The field analysis of an air-filled waveguide can be extended to a dielectric filled waveguide [23], with permeability μ and permittivity ε, and the resulting propagation constant (γ) and wave-impedance (Z) are:

$$\gamma^2 = k_x^2 - \omega^2 \mu\varepsilon \tag{A1}$$

and

$$Z = \frac{j\omega\mu}{\gamma} \tag{A2}$$

where k_x is the mode cut-off wavenumber, which for TE_{m0} modes in a waveguide of width a are:

$$k_x = \frac{m\pi}{a} \tag{A3}$$

where m is a non-zero integer.

For a given mode, a transmission line analogy can be used to represent a section of waveguide [23]. For the case of a uniform cross-section, the propagation constant and characteristic impedance of the transmission line use γ and wave-impedance Z respectively.

For a lossless MNG or ENG, either μ or ε will be negative valued and the other positive, and (A1) shows that γ only takes pure real values and means that only evanescent modes are possible. Equation (A2) indicates that Z will be pure imaginary and capacitive for MNG, and inductive for ENG.

Appendix B

In this appendix, the tunneling conditions derivation are derived. This derivation is based upon voltage-current chain or ABCD parameters [23] noting this problem can that alternatively formulated using wave-transmission parameters [7].

We consider the cascade of a pair of waveguide sections with identical transverse cross sections and therefore the only discontinuity is due to the different materials that fill each of these sections. These two sections could correspond to Sections 2 and 3 of Figure 1 which contain MNG and ENG fills respectively. As both sections are operating in their evanescent modes, they can be described by a pure real propagation constant (being the attenuation constant) and a pure imaginary wave-impedance α_1 and jX_1 respectively for Section 1, α_2 and jX_2 respectively for Section 2. The values of α_1 and α_2 are assumed to take positive values for passive media, whereas the X_1 and X_2 can take positive and negative values. For a given mode, a section of waveguide can be modelled as a transmission line described by its propagation constant and characteristic impedance. Using an ABCD parameter formulation [23], the ABCD matrix of the cascade of the two waveguide sections of lengths l_1 and l_2 is:

$$ABCD_{cascade} = \begin{bmatrix} \cosh(\alpha_1 l_1) & jX_1\sinh(\alpha_1 l_1) \\ \frac{1}{jX_1}\sinh(\alpha_1 l_1) & \cosh(\alpha_1 l_1) \end{bmatrix} \begin{bmatrix} \cosh(\alpha_2 l_2) & jX_2\sinh(\alpha_2 l_2) \\ \frac{1}{jX_2}\sinh(\alpha_2 l_2) & \cosh(\alpha_2 l_2) \end{bmatrix} \tag{A4}$$

Expanding:

$$ABCD_{cascade} = \begin{bmatrix} \cosh(\alpha_1 l_1)\cosh(\alpha_2 l_2) + \frac{X_1}{X_2}\sinh(\alpha_1 l_1)\sinh(\alpha_2 l_2) \\ \frac{1}{jX_2}\sinh(\alpha_2 l_2)\cosh(\alpha_1 l_1) + \frac{1}{jX_1}\sinh(\alpha_1 l_1)\cosh(\alpha_2 l_2) \\ jX_2\sinh(\alpha_2 l_2)\cosh(\alpha_1 l_1) + jX_1\sinh(\alpha_1 l_1)\cosh(\alpha_2 l_2) \\ \cosh(\alpha_1 l_1)\cosh(\alpha_2 l_2) + \frac{X_2}{X_1}\sinh(\alpha_1 l_1)\sinh(\alpha_2 l_2) \end{bmatrix} \tag{A5}$$

An identity ABCD corresponds to perfect transmission with zero reflection. Equation (A5) is equal to the identity matrix when $\alpha_1 l_1 = \alpha_2 l_2$ and $X_1 = -X_2$ which are denoted the attenuation and impedance tunneling conditions respectively. The latter condition requires X_1 and X_2 to be opposite signed and represents resonance across the boundary between Sections 1 and 2 [6,8].

Appendix C

In this appendix, the fields are formulated for TE_{m0} modes in a rectangular waveguide filled with anisotropic dielectrics. The outcome of this analysis is the attenuation constant and wave-impedance for the special case of anisotropic ENG and MNG dielectric fill.

Consider an anisotropic dielectric material whose permeability and permittivity tensors in rectangular coordinates are respectively:

$$\mu = \begin{bmatrix} \mu_{xx} & 0 & 0 \\ 0 & \mu_{yy} & 0 \\ 0 & 0 & \mu_{zz} \end{bmatrix} \tag{A6}$$

$$\varepsilon = \begin{bmatrix} \varepsilon_{xx} & 0 & 0 \\ 0 & \varepsilon_{yy} & 0 \\ 0 & 0 & \varepsilon_{zz} \end{bmatrix} \tag{A7}$$

where μ_{xx}, μ_{yy} and μ_{zz} are the x, y and z directed permeability respectively, and ε_{xx}, ε_{yy} and ε_{zz} are the x, y and z directed permittivity respectively. It is therefore assumed that the material is non-bianisotropic, non-gyrotropic and non-chiral which give zero off-diagonal elements. This assumption is reasonable for the types of metamaterials considered in this work comprising either broad-side-coupled SRRs, or thin strips.

Consider a rectangular waveguide directed in the z direction and whose perfect-electric-conducting walls are located at $x = 0$, $x = a$, $y = 0$ and $y = b$. All field component phasors of a wave propagating in the z direction will contain the common factor $\exp(-\gamma z)$ where γ is the complex valued propagation constant. If we restrict the waves to be TE (transverse electric) waves, and further restrict the wave to containing no y dependence (i.e., TE_{m0} modes), substitution of (A6) and (A7) into Maxwell's curl equations in phasor form, yields:

$$\gamma E_y = -j\omega\mu_{xx}H_x \tag{A8a}$$

$$\frac{\partial E_y}{\partial x} = -j\omega\mu_{zz}H_z \tag{A8b}$$

$$-\gamma H_x - \frac{\partial H_z}{\partial x} = j\omega\varepsilon_{yy}E_y \tag{A8C}$$

Equations (A8a)–(A8c) shows that only the tensor components μ_{xx}, ε_{yy} and μ_{zz} play a role in TE_{m0} mode behavior. Based upon the waveguide boundary conditions, the solution of the non-zero field phasors, that also satisfy (A8a) and (A8b) must take the form:

$$H_x = \frac{-\gamma A}{j\omega\mu_{xx}}\sin k_x x \tag{A9a}$$

$$E_y = A\sin k_x x \tag{A9b}$$

$$H_z = \frac{-k_x A}{j\omega\mu_{zz}}\cos k_x x \tag{A9c}$$

where the cut-off wavenumber k_x is given by (A3), A is an amplitude (which could be complex valued), and m is a non-zero integer. Substituting (A9) into (A8c) yields the dispersion relation:

$$\gamma^2 = \frac{\mu_{xx}}{\mu_{zz}}k_x{}^2 - \omega^2\mu_{xx}\varepsilon_{yy} \tag{A10}$$

For an evanescent mode, γ will be pure real and equal to the attenuation constant α which is positive valued. For a TE wave travelling (or decaying) in the z direction, the wave-impedance is $-E_y/H_x$. Hence from (A9a) and (A9b):

$$Z_{TE} = \frac{j\omega\mu_{xx}}{\alpha} \tag{A11}$$

Thus from (A11), Z_{TE} will be pure imaginary whose is sign is that of μ_{xx}.

There are two types of anisotropic dielectric that are of interest here. They are MNG with negative permeability confined to the x direction, and ENG with negative permittivity confined to the y direction.

C.1. Anisotropic MNG

For the first case, μ_{xx} is negative valued, μ_{zz} equals μ_0 the permeability of a vacuum, and ε_{yy} is positive valued. The permittivity ε_{yy} is greater than the permittivity of a free-space (ε_0) due to the inherent electric polarizability of the constituent elements comprising the metamaterial, and the presence of insulating substrates to support such elements. Hence, from (A10):

$$\alpha = \sqrt{\frac{-\mu_{xx}}{\mu_0}\left(\omega^2\mu_0\varepsilon_{yy} - k_x{}^2\right)} \tag{A12}$$

and is real (i.e., evanescent mode) only if $\omega^2\mu_0\varepsilon_{yy} > k_x{}^2$ which is denoted the "anti-cut-off" property [7]. Clearly from (A11), Z_{TE} will be a capacitive reactance.

C.2. Anisotropic ENG

For the second case, $\mu_{xx} = \mu_{zz} = \mu_0$, and ε_{yy} is negative valued. Hence, from (A10):

$$\alpha = \sqrt{k_x{}^2 - \omega^2 \mu_0 \varepsilon_{yy}} \tag{A13}$$

and is real for all frequencies provided ε_{yy} is negative valued. Clearly from (A11), Z_{TE} will be an inductive reactance.

The foregoing indicates that pairing of the above anisotropic ENG and MNG filled waveguides can yield the required impedance tunneling condition. Namely, apart from the anti-cut-off property, the conclusions are similar to the isotropic ENG and MNG cases.

In practice, for SRR realizations of anisotropic MNG, and strip realization of ENG, μ_{xx} and ε_{yy} are frequency dependent and are only negative over a restricted frequency range; particularly for the SRR loaded waveguide. Therefore, the above theoretical analysis needs to be interpreted in that context.

References

1. Silveirinha, M.; Engheta, N. Tunneling of electromagnetic energy through subwavelength channels and bends using ε-near-zero materials. *Phys. Rev. Lett.* **2006**, *97*, 157403. [CrossRef] [PubMed]
2. Liu, R.; Cheng, Q.; Hand, T.; Mock, J.J.; Cui, T.J.; Cummer, S.A.; Smith, D.R. Experimental demonstration of electromagnetic tunneling through an epsilon-near-zero metamaterial at microwave frequencies. *Phys. Rev. Lett.* **2008**, *100*, 023903. [CrossRef] [PubMed]
3. Edwards, B.; Alù, A.; Young, M.E.; Silveirinha, M.; Engheta, N. Experimental verification of epsilon-near-zero metamaterial coupling and energy squeezing using a microwave waveguide. *Phys. Rev. Lett.* **2008**, *100*, 033903. [CrossRef] [PubMed]
4. Mitrovic, M.; Jokanovic, B.; Vojnovic, N. Wideband tuning of the tunneling frequency in a narrowed epsilon-near-zero channel. *IEEE Antennas Wirel. Propag. Lett.* **2013**, *12*, 631–634. [CrossRef]
5. Soric, J.C.; Alù, A. Longitudinally independent matching and arbitrary wave patterning using ε-near-zero channels. *IEEE Trans. Microw. Theory Tech.* **2015**, *63*, 3558–3567. [CrossRef]
6. Alù, A.; Engheta, N. Pairing an epsilon-negative slab with a mu-negative slab: Resonance, tunneling and transparency. *IEEE Trans. Antennas Propag.* **2003**, *51*, 2558–2571. [CrossRef]
7. Baena, J.D.; Jelinek, L.; Marqués, R.; Medina, F. Near-perfect tunneling and amplification of evanescent electromagnetic waves in a waveguide filled by a metamaterial: Theory and experiments. *Phys. Rev. B* **2005**, *72*, 075116. [CrossRef]
8. Liu, R.; Zhao, B.; Lin, X.Q.; Cheng, Q.; Cui, T.J. Evanescent-wave amplification studied using a bilayer periodic circuit structure and its effective medium model. *Phys. Rev. B* **2007**, *75*, 125118. [CrossRef]
9. Feng, Y.; Teng, X.; Wang, Z.; Zhao, J.; Jiang, T. Extraordinary transmission with evanescent wave enhancement in planar waveguide loaded with anisotropic metamaterials. In Proceedings of the 2008 Asia–Pacific Microwave Conference, Hong Kong/Macao, China, 16–20 December 2008; IEEE: New York, NY, USA, 2008. [CrossRef]
10. Feng, T.; Li, Y.; Jiang, H.; Sun, Y.; He, L.; Li, H.; Zhang, Y.; Shi, Y.; Chen, H. Electromagnetic tunneling in a sandwich structure containing single negative media. *Phys. Rev. E* **2009**, *79*, 026601. [CrossRef]
11. Liu, C.H.; Behdad, N. Analysis of electromagnetic wave tunneling through stacked single-negative metamaterial slabs: A microwave filter theory approach. In Proceedings of the 2012 IEEE International Symposium on Antennas and Propagation, Chicago, IL, USA, 8–14 July 2012; IEEE: New York, NY, USA, 2012. [CrossRef]
12. Zhang, L.; Wang, J.; He, L.; Qiao, W.; Chen, L.; Wang, Q.; Zhao, Y. Tunneling time in a conjugate matched pair consisting of ε-negative and μ-negative materials. *Physica B* **2013**, *431*, 127–131. [CrossRef]
13. Zhang, L.; Zhang, Y.; Chen, X.; Liu, X.; Zhang, L.; Chen, H. Study on the tunneling mode in a sub-wavelength open-cavity resonator consisting of single negative materials. *IEEE Trans. Antennas Propag.* **2014**, *62*, 504–508. [CrossRef]

14. Zhao, J.; Zhang, Y.; Yu, X.; Li, Y.; He, L.; Fang, K. Electromagnetic tunneling through epsilon-negative waveguide paired with mu-negative waveguide. In Proceedings of the 2014 Asia-Pacific Conference on Antennas and Propagation, Harbin, China, 26–29 July 2014; IEEE: New York, NY, USA, 2014. [CrossRef]

15. Chen, Y.; Huang, S.; Yan, X.; Shi, J. Electromagnetic tunneling through conjugated single-negative metamaterial pairs and double-positive layer with high-magnetic fields. *Chin. Opt. Lett.* **2014**, *12*, 101601. [CrossRef]

16. Marques, R.; Martel, J.; Mesa, F.; Medina, F. Left-handed-media simulation and transmission of EM waves in subwavelength split-ring-resonator-loaded metallic waveguides. *Phys. Rev. Lett.* **2002**, *89*, 183901. [CrossRef]

17. Hrabar, S.; Bartolic, J.; Sipus, Z. Waveguide miniaturization using uniaxial negative permeability metamaterial. *IEEE Trans. Antennas Propag.* **2005**, *53*, 110–119. [CrossRef]

18. Odabasi, H.; Teixeira, F.L. Electric-field-coupled resonators as metamaterial loadings for waveguide miniaturization. *J. Appl. Phys.* **2013**, *114*, 214901. [CrossRef]

19. Pollock, J.G.; Iyer, A.K. Below-cutoff propagation in metamaterial-lined circular waveguides. *IEEE Trans. Microw. Theory Tech.* **2013**, *61*, 3169–3178. [CrossRef]

20. Esteban, J.; Camacho-Peñalosa, C.; Page, J.E.; Martín-Guerrero, T.M.; Márquez-Segura, E. Simulation of negative permittivity and negative permeability by means of evanescent waveguide modes—Theory and experiment. *IEEE Trans. Microw. Theory Tech.* **2005**, *53*, 1506–1514. [CrossRef]

21. Pendry, J.B.; Holden, A.J.; Robbins, D.J.; Stewart, W.J. Low frequency plasmons in thin-wire structures. *J. Phys. Condens. Matter* **1998**, *10*, 4785–4809. [CrossRef]

22. Pendry, J.B.; Holden, A.J.; Robbins, D.J.; Stewart, W.J. Magnetism from conductors and enhanced nonlinear phenomena. *IEEE Trans. Microw. Theory Tech.* **1999**, *47*, 2075–2084. [CrossRef]

23. Ramo, S.; Whinnery, J.R.; van Duzer, T. *Fields and Waves in Communication Electronics*; John Wiley & Sons, Inc.: New York, NY, USA, 1994; ISBN 0-471-58551-34.

24. Eccleston, K.W.; Platt, I.G. Printed NRI metamaterial based on broad-side-coupled SRRs and a thinned array of strips. In Proceedings of the 2016 Asia-Pacific Microwave Conference, New Delhi, India, 5–9 December 2016; IEEE: New York, NY, USA, 2017. [CrossRef]

25. Smith, D.R.; Padilla, W.J.; Vier, D.C.; Nemat-Nasser, S.C.; Schultz, S. Composite medium with simultaneously negative permeability and permittivity. *Phys. Rev. Lett.* **2000**, *84*, 4184–4187. [CrossRef]

26. Pozar, D.M. *Microwave Engineering*, 3rd ed.; John Wiley & Sons Inc.: Hoboken, NJ, USA, 2005; ISBN 978-0-470-63155-3.

27. Eberspächer, M.A.; Eibert, T.F. Extraction of embedded dispersion characteristics. In Proceedings of the 2011 Asia-Pacific Microwave Conferenc, Melbourne, Australia, 5–8 December 2011; IEEE: New York, NY, USA, 2011; pp. 801–804.

28. Williams, D.F.; Jargon, J.; Arz, U.; Hale, P. Rectangular-waveguide impedance. In Proceedings of the 85th ARFTG Conference, Phoenix, AZ, USA, 22 May 2015; IEEE: New York, NY, USA, 2015. [CrossRef]

29. Marqués, R.; Mesa, F.; Martel, J.; Medina, F. Comparative analysis of edge-and broadside-coupled split ring resonators for metamaterial design-theory and experiments. *IEEE Trans. Antennas Propag.* **2003**, *51*, 2572–2581. [CrossRef]

30. Eccleston, K.W. A new interpretation of through-line deembedding. *IEEE Trans. Microw. Theory Tech.* **2016**, *64*, 3887–3893. [CrossRef]

31. Baskey, H.B.; Akhtar, M.J. An improved measurement technique for retrieval of effective constitutive properties of thin dielectric/magnetic and metamaterial samples. In Proceedings of the 83rd ARFTG Microwave Measurement Conference, Tampa, FL, USA, 6 June 2014; IEEE: New York, NY, USA, 2014. [CrossRef]

32. Hrabar, S.; Benic, L.; Bartolic, J. Simple experimental determination of complex permittivity or complex permeability of SNG metamaterials. In Proceedings of the 36th European Microwave Conference, Manchester, UK, 10–15 September 2006; IEEE: New York, NY, USA, 2006; pp. 1395–1398. [CrossRef]

33. Chen, H.; Zhang, J.; Bai, Y.; Luo, Y.; Ran, L.; Jiang, Q.; Kong, J.A. Experimental retrieval of the effective parameters of metamaterials based on a waveguide method. *Opt. Express* **2006**, *14*, 12944–12949. [CrossRef] [PubMed]

34. Kim, S.; Kuester, E.F.; Holloway, C.L.; Scher, A.D.; Baker-Jarvis, J. Boundary effects on the determination of metamaterial parameters from normal incidence reflection and transmission measurements. *IEEE Trans. Antennas Propag.* **2011**, *59*, 2226–2240. [CrossRef]

35. Arslanagić, S.; Hansen, T.V.; Mortensen, N.A.; Gregersen, A.H.; Sigmund, O.; Ziolkowski, R.W.; Breinbjerg, O. A review of the scattering-parameter extraction method with clarification of ambiguity issues in relation to metamaterial homogenization. *IEEE Antennas Propag. Mag.* **2013**, *55*, 91–106. [CrossRef]
36. Machac, J.; Jelinek, L. Metamaterial made of BC-SRRs with randomly dispersed resonance frequencies. In Proceedings of the 2015 IEEE MTT-S International Microwave Symposium, Phoenix, AZ, USA, 17–22 May 2015; IEEE: New York, NY, USA, 2015. [CrossRef]
37. Gonzalez, G. *Microwave Transistor Amplifiers—Analysis and Design*, 2nd ed.; Prentice-Hall: Upper Saddle River, NJ, USA, 1997; ISBN 0-13-254335-4.

MDPI

St. Alban-Anlage 66

4052 Basel

Switzerland

Tel. +41 61 683 77 34

Fax +41 61 302 89 18

www.mdpi.com

Electronics Editorial Office

E-mail: electronics@mdpi.com

www.mdpi.com/journal/electronics